Math Challenge II-B

Number Theory

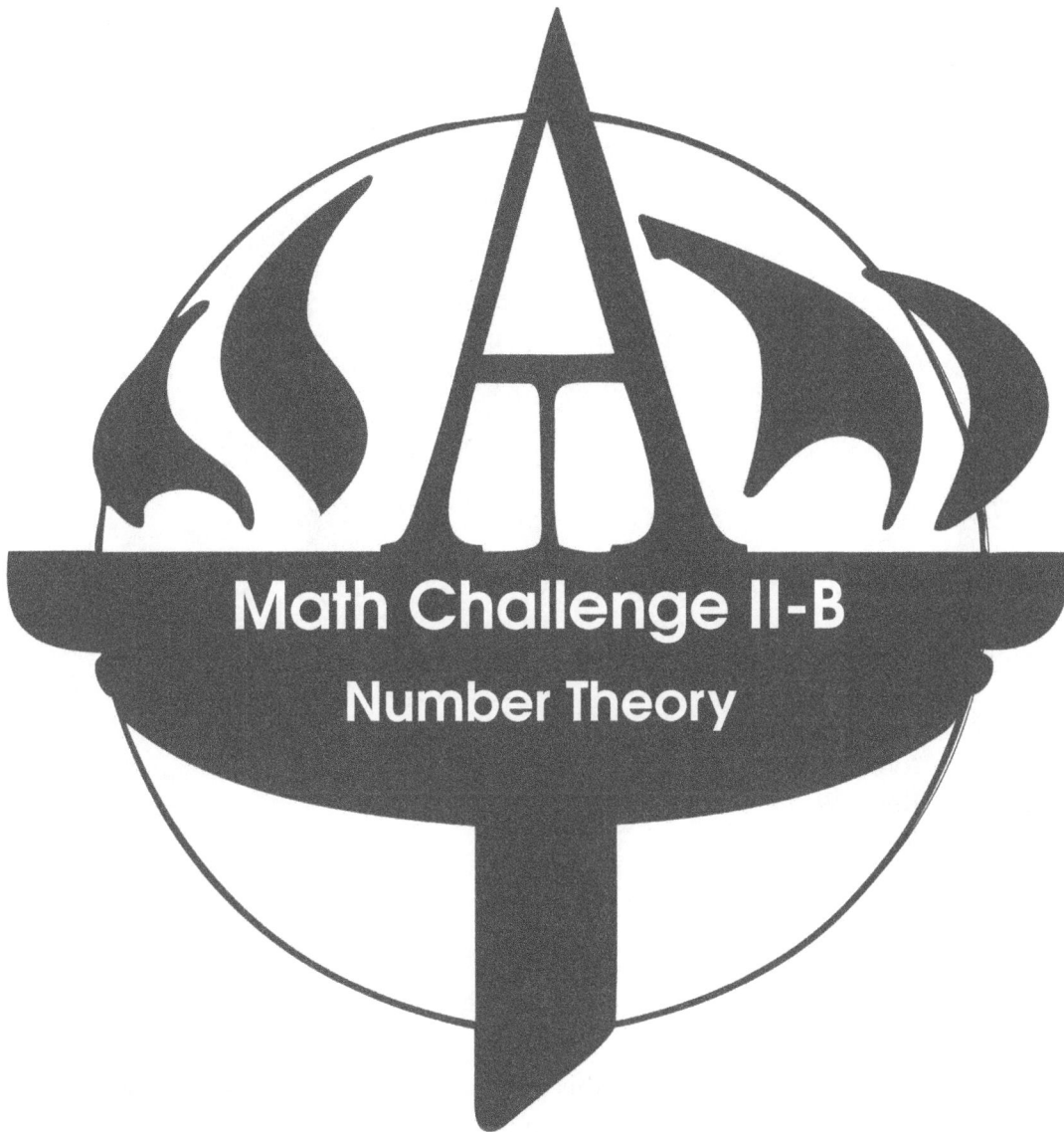

Areteem Institute

Math Challenge II-B Number Theory

Series: Math Challenge Curriculum Textbooks, Vol. 23

Edited by   David Reynoso
             John Lensmire
             Kevin Wang
             Kelly Ren

ISBN: 1-944863-42-7
ISBN-13: 978-1-944863-42-5
First printing, February 2019.

Math Challenge II-A Combinatorics
Math Challenge II-B Combinatorics
Math Challenge III Combinatorics
Math Challenge I-A Number Theory
Math Challenge I-B Number Theory
Math Challenge I-C Finite Math
Math Challenge II-A Number Theory
Math Challenge II-B Number Theory
Math Challenge III Number Theory

COMING SOON FROM ARETEEM PRESS

Fun Math Problem Solving For Elementary School Vol. 2 (and Solutions Manual)
Counting & Probability for Middle School (and Solutions Manual) - From Common
Core to Math Competitions
Number Theory Problem Solving for Middle School (and Solutions Manual) -
From Common Core to Math Competitions

The books are available in paperback and eBook formats (including Kindle and other
formats).
To order the books, visit https://areteem.org/bookstore.

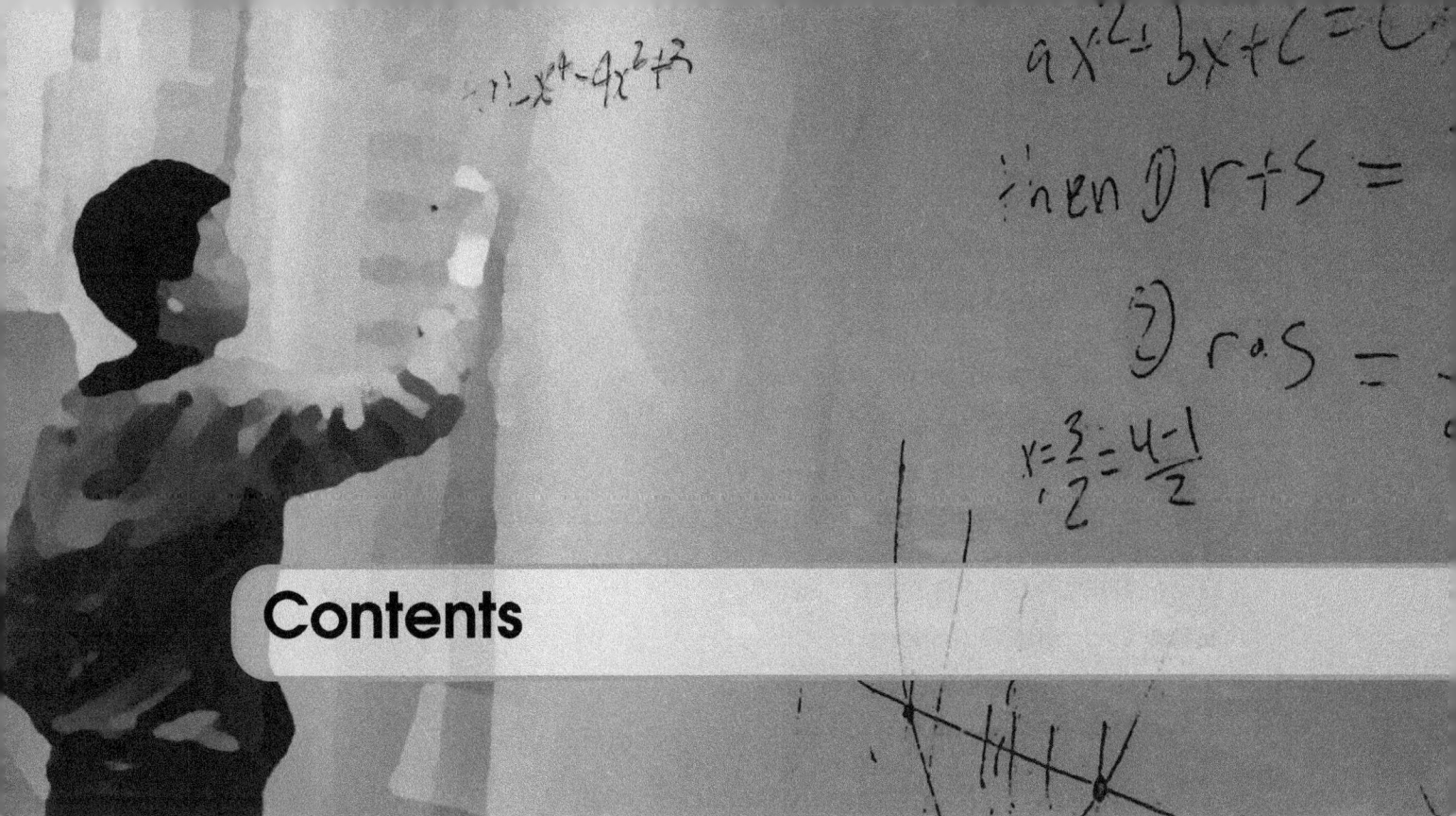

Contents

Introduction 9

1   Place Values and Number Bases 17
1.1   Example Questions 18
1.2   Quick Response Questions 22
1.3   Practice Questions 24

2   Divisibility 27
2.1   Example Questions 28
2.2   Quick Response Questions 31
2.3   Practice Questions 33

3   Prime Numbers 35
3.1   Example Questions 37
3.2   Quick Response Questions 40
3.3   Practice Questions 42

4   Modular Arithmetic 45
4.1   Example Questions 45
4.2   Quick Response Questions 49

4.3   Practice Questions 51

5   Problem Solving in Mod. Arith. 53
5.1   Example Questions 53
5.2   Quick Response Questions 57
5.3   Practice Questions 59

6   Theorems in Mod. Arith. 61
6.1   Example Questions 62
6.2   Quick Response Questions 65
6.3   Practice Questions 67

7   Advanced Theorems in Mod. Arith. 69
7.1   Example Questions 71
7.2   Quick Response Questions 75
7.3   Practice Questions 77

8   Diophantine Equations 79
8.1   Example Questions 80
8.2   Quick Response Questions 82
8.3   Practice Questions 84

9   Floor Function 87
9.1   Example Questions 88
9.2   Quick Response Questions 91
9.3   Practice Questions 93

Solutions to the Example Questions 95
1   Solutions to Chapter 1 Examples 96
2   Solutions to Chapter 2 Examples 104
3   Solutions to Chapter 3 Examples 109
4   Solutions to Chapter 4 Examples 115
5   Solutions to Chapter 5 Examples 121
6   Solutions to Chapter 6 Examples 127

7   **Solutions to Chapter 7 Examples** 132

8   **Solutions to Chapter 8 Examples** 139

9   **Solutions to Chapter 9 Examples** 145

Introduction

The math challenge curriculum textbook series is designed to help students learn the fundamental mathematical concepts and practice their in-depth problem solving skills with selected exercise problems. Ideally, these textbooks are used together with Areteem Institute's corresponding courses, either taken as live classes or as self-paced classes. According to the experience levels of the students in mathematics, the following courses are offered:

- Fun Math Problem Solving for Elementary School (grades 3-5)
- Algebra Readiness (grade 5; preparing for middle school)
- Math Challenge I-A Series (grades 6-8; intro to problem solving)
- Math Challenge I-B Series (grades 6-8; intro to math contests e.g. AMC 8, ZIML Div M)
- Math Challenge I-C Series (grades 6-8; topics bridging middle and high schools)
- Math Challenge II-A Series (grades 9+ or younger students preparing for AMC 10)
- Math Challenge II-B Series (grades 9+ or younger students preparing for AMC 12)
- Math Challenge III Series (preparing for AIME, ZIML Varsity, or equivalent contests)
- Math Challenge IV Series (Math Olympiad level problem solving)

These courses are designed and developed by educational experts and industry professionals to bring real world applications into the STEM education. These programs are ideal for students who wish to win in Math Competitions (AMC, AIME, USAMO, IMO,

ARML, MathCounts, Math League, Math Olympiad, ZIML, etc.), Science Fairs (County Science Fairs, State Science Fairs, national programs like Intel Science and Engineering Fair, etc.) and Science Olympiad, or purely want to enrich their academic lives by taking more challenges and developing outstanding analytical, logical thinking and creative problem solving skills.

In Math Challenge II-B, students learn and practice in areas such as algebra and geometry at the high school level, as well as advanced number theory and combinatorics. Topics include polynomials, inequalities, special algebraic techniques, trigonometry, triangles and polygons, collinearity and concurrency, vectors and coordinates, numbers and divisibility, modular arithmetic, residue classes, advanced counting strategies, binomial coefficients, and various other topics and problem solving techniques involved in math contests such as the American Mathematics Competition (AMC) 10 & 12, ARML, beginning AIME, and Zoom International Math League (ZIML) Junior Varsity and Varsity Divisions.

The course is divided into four terms:

- Summer, covering Algebra
- Fall, covering Geometry
- Winter, covering Combinatorics
- Spring, covering Number Theory

The book contains course materials for Math Challenge II-B: Number Theory.

We recommend that students take all four terms. Each of the individual terms is self-contained and does not depend on other terms, so they do not need to be taken in order, and students can take single terms if they want to focus on specific topics.

Students can sign up for the live or self-paced course at `classes.areteem.org`.

About Areteem Institute

Areteem Institute is an educational institution that develops and provides in-depth and advanced math and science programs for K-12 (Elementary School, Middle School, and High School) students and teachers. Areteem programs are accredited supplementary programs by the Western Association of Schools and Colleges (WASC). Students may attend the Areteem Institute in one or more of the following options:

- Live and real-time face-to-face online classes with audio, video, interactive online whiteboard, and text chatting capabilities;
- Self-paced classes by watching the recordings of the live classes;
- Short video courses for trending math, science, technology, engineering, English, and social studies topics;
- Summer Intensive Camps held on prestigious university campuses and Winter Boot Camps;
- Practice with selected free daily problems and monthly ZIML competitions at ziml.areteem.org.

Areteem courses are designed and developed by educational experts and industry professionals to bring real world applications into STEM education. The programs are ideal for students who wish to build their mathematical strength in order to excel academically and eventually win in Math Competitions (AMC, AIME, USAMO, IMO, ARML, MathCounts, Math Olympiad, ZIML, and other math leagues and tournaments, etc.), Science Fairs (County Science Fairs, State Science Fairs, national programs like Intel Science and Engineering Fair, etc.) and Science Olympiads, or for students who purely want to enrich their academic lives by taking more challenging courses and developing outstanding analytical, logical, and creative problem solving skills.

Since 2004 Areteem Institute has been teaching with methodology that is highly promoted by the new Common Core State Standards: stressing the conceptual level understanding of the math concepts, problem solving techniques, and solving problems with real world applications. With the guidance from experienced and passionate professors, students are motivated to explore concepts deeper by identifying an interesting problem, researching it, analyzing it, and using a critical thinking approach to come up with multiple solutions.

Thousands of math students who have been trained at Areteem have achieved top honors and earned top awards in major national and international math competitions, including Gold Medalists in the International Math Olympiad (IMO), top winners and qualifiers at the USA Math Olympiad (USAMO/JMO) and AIME, top winners at the

Zoom International Math League (ZIML), and top winners at the MathCounts National Competition. Many Areteem Alumni have graduated from high school and gone on to enter their dream colleges such as MIT, Cal Tech, Harvard, Stanford, Yale, Princeton, U Penn, Harvey Mudd College, UC Berkeley, or UCLA. Those who have graduated from colleges are now playing important roles in their fields of endeavor.

Further information about Areteem Institute, as well as updates and errata of this book, can be found online at http://www.areteem.org.

About Zoom International Math League

The Zoom International Math League (ZIML) has a simple goal: provide a platform for students to build and share their passion for math and other STEM fields with students from around the globe. Started in 2008 as the Southern California Mathematical Olympiad, ZIML has a rich history of past participants who have advanced to top tier colleges and prestigious math competitions, including American Math Competitions, MATHCOUNTS, and the International Math Olympaid.

The ZIML Core Online Programs, most available with a free account at ziml.areteem.org, include:

- **Daily Magic Spells:** Provides a problem a day (Monday through Friday) for students to practice, with full solutions available the next day.
- **Weekly Brain Potions:** Provides one problem per week posted in the online discussion forum at ziml.areteem.org. Usually the problem does not have a simple answer, and students can join the discussion to share their thoughts regarding the scenarios described in the problem, explore the math concepts behind the problem, give solutions, and also ask further questions.
- **Monthly Contests:** The ZIML Monthly Contests are held the first weekend of each month during the school year (October through June). Students can compete in one of 5 divisions to test their knowledge and determine their strengths and weaknesses, with winners announced after the competition.
- **Math Competition Practice:** The Practice page contains sample ZIML contests and an archive of AMC-series tests for online practice. The practices simulate the real contest environment with time-limits of the contests automatically controlled by the server.
- **Online Discussion Forum:** The Online Discussion Forum is open for any comments and questions. Other discussions, such as hard Daily Magic Spells or the Weekly Brain Potions are also posted here.

These programs encourage students to participate consistently, so they can track their progress and improvement each year.

In addition to the online programs, ZIML also hosts onsite Local Tournaments and Workshops in various locations in the United States. Each summer, there are onsite ZIML Competitions at held at Areteem Summer Programs, including the National ZIML Convention, which is a two day convention with one day of workshops and one day of competition.

ZIML Monthly Contests are organized into five divisions ranging from upper elementary school to advanced material based on high school math.

- **Varsity:** This is the top division. It covers material on the level of the last 10 questions on the AMC 12 and AIME level. This division is open to all age levels.
- **Junior Varsity:** This is the second highest competition division. It covers material at the AMC 10/12 level and State and National MathCounts level. This division is open to all age levels.
- **Division H:** This division focuses on material from a standard high school curriculum. It covers topics up to and including pre-calculus. This division will serve as excellent practice for students preparing for the math portions of the SAT or ACT. This division is open to all age levels.
- **Division M:** This division focuses on problem solving using math concepts from a standard middle school math curriculum. It covers material at the level of AMC 8 and School or Chapter MathCounts. This division is open to all students who have not started grade 9.
- **Division E:** This division focuses on advanced problem solving with mathematical concepts from upper elementary school. It covers material at a level comparable to MOEMS Division E. This division is open to all students who have not started grade 6.

The ZIML site features are also provided on the ZIML Mobile App, which is available for download from Apple's App Store and Google Play Store.

Acknowledgments

This book contains many years of collaborative work by the staff of Areteem Institute. This book could not have existed without their efforts. Huge thanks go to the Areteem staff for their contributions!

The examples and problems in this book were either created by the Areteem staff or adapted from various sources, including other books and online resources. Especially, some good problems from previous math competitions and contests such as AMC, AIME, ARML, MATHCOUNTS, and ZIML are chosen as examples to illustrate concepts or problem-solving techniques. The original resources are credited whenever possible. However, it is not practical to list all such resources. We extend our gratitude to the original authors of all these resources.

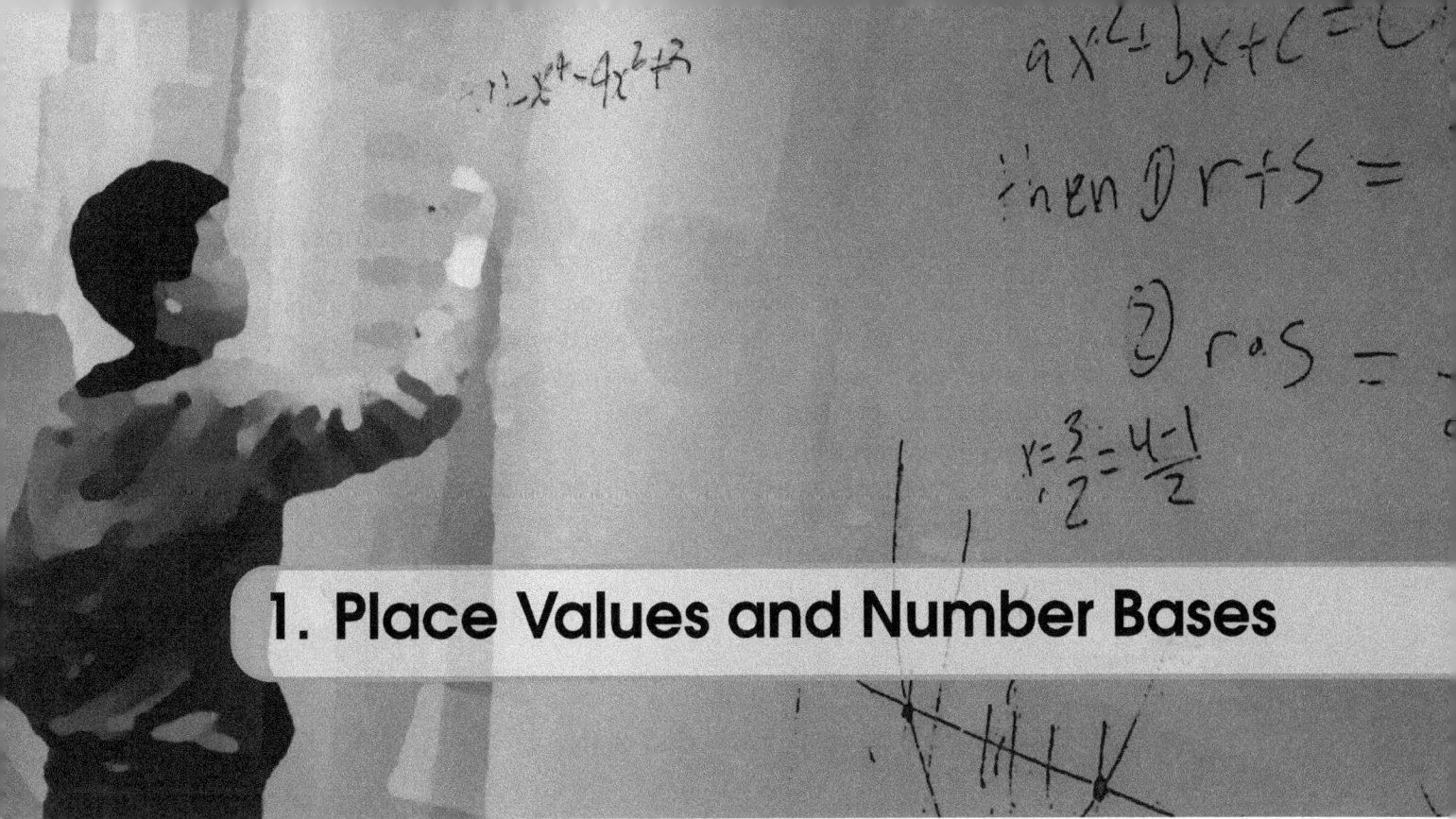

1. Place Values and Number Bases

Place Values

- The value of a digit depends on its place, or position in a number. For example,

$$654 = 6 \times 100 + 5 \times 10 + 4 \times 1$$
$$2016 = 2 \times 1000 + 0 \times 100 + 1 \times 10 + 6 \times 1$$

- In general,

$$\overline{abc} = a \times 100 + b \times 10 + c$$
$$\overline{a_n a_{n-1} \cdots a_0} = a_n \times 10^n + a_{n-1} \times 10^{n-1} + \cdots + a_0$$

- This can be extended to non-integers as well. For example

$$2.71 = 2 \times 1 + 7 \times \frac{1}{10} + 1 \times \frac{1}{100}.$$

- This system of expressing numbers is referred to as the *decimal* or *base* 10 system. It uses the digits $0, 1, 2, 3, \ldots, 9$.

Other Bases

- Other positive integers (other than 1) can be used as bases as well. Numbers in these bases are expressed in a similar manner as the decimal system.
- For example, the *binary* or base 2 system, uses two symbols (0 and 1) to represent numbers using place values that are powers of two (1, 2, 4, etc.).

- For bases other than 10, the base number is usually written as subscripts. For example, 101_2 would be the binary number with digits $1, 0, 1$.
- Given a number expressed in other bases, we can find the equivalent values in decimal by adding up the place values.
- Some common bases include:
  - Binary (base 2) numbers with digits $0, 1$. For example,
  $$1010_2 = 1 \times 2^3 + 0 \times 2^2 + 1 \times 2 + 0 = 10.$$
  - Octal (base 8) numbers with digits $0, 1, \ldots, 7$. For example,
  $$654_8 = 6 \times 8^2 + 5 \times 8 + 4 = 428.$$
  - Hexadecimal (base 16) numbers with digits $0, 1, \ldots, 9, A, B, C, D, E, F$. For example,
  $$ABC123_{16} = 10 \times 16^5 + 11 \times 16^4 + 12 \times 16^3 + 1 \times 16^2 + 2 \times 16 + 3 = 11256099.$$
- Also note that operations (such as $+, \times$) work the same way as in the decimal system. Just keep in mind that you "carry" at the base number (for example, in the binary system, carry occurs whenever you have "2").
- For example, $10 + 10 = 20$ in base 10. In base 2, $10 = 1010_2$. Adding in base 2 we have
$$\begin{array}{r} 1\ 0\ 1\ 0 \\ +\ 1\ 0\ 1\ 0 \\ \hline 1\ 0\ 1\ 0\ 0 \end{array}$$
where $10100_2 = 16 + 4 = 20$ as expected.

1.1 Example Questions

Problem 1.1 Convert the following numbers written in different bases to decimal.

(a) 1232_4

(b) 10120_3

(c) ABC_{16}

Problem 1.2 Convert the number 98 in decimal to the following bases:

(a) 7

(b) 3

(c) 2

(d) 16

Problem 1.3 Convert the following. Try to come up with an efficient strategy!

(a) 1234567890 in base 10 to base 100. Note: Think of base 100 as having digits $\boxed{00}$, $\boxed{01}$, ..., $\boxed{99}$.

(b) 1011010110_2 to base 8.

(c) 4095 from decimal to hexadecimal

(d) 1333 from octal to ternary.

Problem 1.4 Perform the following operations in the given base. Double check your answer by converting to and from decimal as well.

(a) $122_3 + 201_3$.

(b) $1010_2 \times 11_2$.

Problem 1.5 The following is a procedure for converting from decimal to binary: Starting with a number N, repeatedly divide by two and record the remainder (0 or 1) at each step. After reaching 0, the remainders written in reverse order will give the number in base 2. For example,

$$10 \div 2 = 5 \ R \ 0; \ 5 \div 2 = 2 \ R \ 1; \ 2 \div 2 = 1 \ R \ 0; \ 1 \div 2 = 0 \ R \ 1 \text{ so } 10 = 1010_2.$$

(a) Use the procedure to convert 47 from decimal to binary.

(b) Explain why the procedure works. Hint: You do not need to give a full proof, but at least explain a small example.

Problem 1.6 Recall $0.375 = 1/4 + 1/8$ so $0.375 = 0.011_2$. Devise a procedure for converting decimals less than 1 to binary. Use your method to convert the following to binary.

(a) 0.3125.

(b) $0.\overline{3}$.

Problem 1.7 Mystery Bases

(a) Find the value of base b such that the following addition is correct:

$$531_b + 135_b = 1110_b.$$

(b) Explain the joke "Halloween equals Christmas".

Problem 1.8 In the sequence $111, 21, x, 12, y, 10, z, \ldots$ what are x, y, z?

Problem 1.9 Based on the digits, how do you tell if a number is even or odd,

(a) if the number is written in base 10?

(b) if the number is written in base 2?

(c) if the number is written in base 3?

Problem 1.10 What is the minimum number of weights which enable us to weigh any integer number of grams of gold from 1 to 100 on a standard balance with two pans? You are allowed to use the weights on either side of the pan.

1.2 Quick Response Questions

Problem 1.11 Convert $9AF_{16}$ to decimal.

Problem 1.12 Convert 1111111_2 (seven 1's) to decimal.

Problem 1.13 Write 345 in octal.

Problem 1.14 Write 1170 in hexadecimal.

Problem 1.15 Convert 10100011_2 to octal.

Problem 1.16 What is $100_2 \cdot 10101_2$? Give your answer in base 2.

Problem 1.17 What is $123_7 + 456_7$? Give your answer in base 7.

Problem 1.18 What is 23_9 squared? Give your answer in base 9.

Problem 1.19 What is $444_5 + 44_5$? Give your answer in base 5.

Problem 1.20 Convert 0.375 to a decimal written in binary.

1.3 Practice Questions

Problem 1.21 Convert the following to decimal.

(a) 414_5

(b) 255 in octal.

Problem 1.22 Convert

(a) 123_5 to base 3.

(b) 190 in decimal to base hexadecimal.

Problem 1.23 Perform the following conversions.

(a) Convert 10100011_2 to base 16.

(b) $12AB_{16}$ to binary.

Problem 1.24 Calculate

(a) $21012_3 + 121201_3$ and give your answer in ternary.

(b) $4096_{10} \times 512_{10}$ and give your answer in hexadecimal.

Problem 1.25 In a previous problem we saw an efficient way to convert from decimal to binary. Can this method be extended to work with other bases as well? Give at least 2 examples to explain your answer. If the answer is yes, explain the general method.

Problem 1.26 In a previous problem we saw an efficient way to convert fractions written as decimals in base 10 to binary. Can this method be extended to work with other bases as well? Give at least 2 examples to explain your answer. If the answer is yes, explain the general method.

Problem 1.27 Is it possible for the following statements to be true in some base?

(a) $3 + 4 = 10$ and $3 \times 4 = 15$.

(b) $2 + 3 = 5$ and $2 \times 3 = 11$.

Problem 1.28 Find the next term in the sequence $10, 13, 16, 22, 31, \ldots$.

Problem 1.29 Give a general procedure to tell whether a number written in base B is even or odd. Your method should extend the example problem where we found rules for $B = 10, 2, 3$. Explain your answer.

Problem 1.30 What is the minimum number of weights which enable us to weight any integer number of grams of gold from 1 to 1000 on a standard balance with two pans if you are only allowed to use the weights on one side of the pan? Explain your answer using number bases.

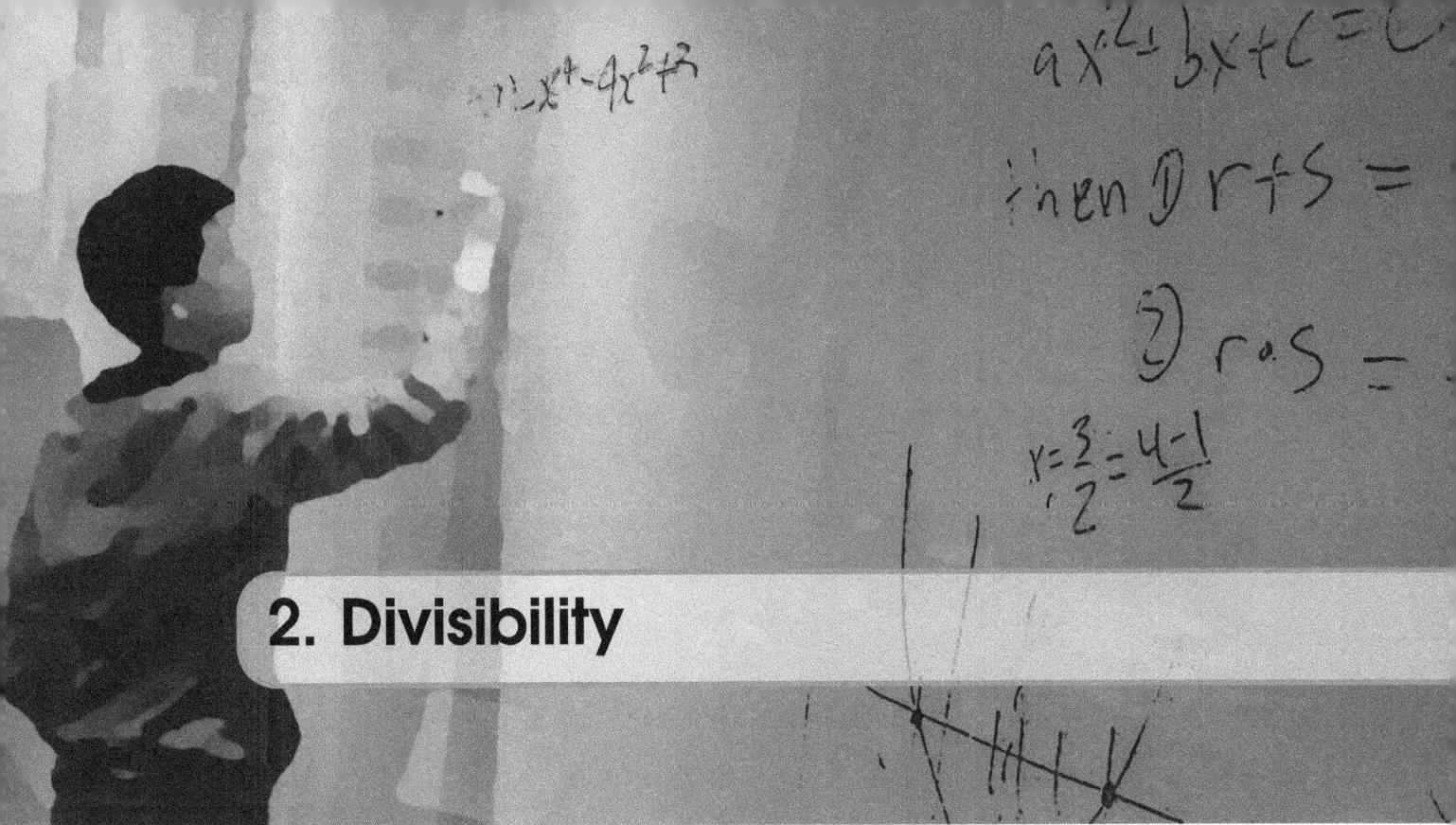

2. Divisibility

Divisibility

- Recall if an integer a divides an integer n evenly (that is, there is an integer k such that $a \cdot k = n$), then a is a factor (or divisor) of n. We write $a \mid n$ to denote that a divides n.
- The following (easy) facts are sometimes useful:
    - If $a \mid n$ and $a \mid m$ then $a \mid n \pm m$.
    - If $a \mid n$ and k is any integer, then $a \mid k \cdot n$.
- If p, q are primes and $p \mid n$ and $q \mid n$ then $p \cdot q \mid n$. We will present the general version of this statement next time.

Divisibility Rules

Let $n = \overline{a_k a_{k-1} \ldots a_1 a_0} = a_k 10^k + a_{k-1} 10^{k-1} + \cdots + a_1 10 + a_0, a_i \in \{0, 1, \ldots, 9\}$.

- By 2^j : If $\overline{a_{j-1} a_{j-2} \ldots a_1 a_0}$ is divisible by 2^j, then $2^j \mid n$.
- By 3: If $3 \mid (a_k + a_{k-1} + \cdots + a_1 + a_0)$, then $3 \mid n$.
- By 5: If $a_0 = 0$ or $a_0 = 5$, then $5 \mid n$.
- By 7: If $7 \mid \overline{a_k a_{k-1} \ldots a_1} - 2a_0$, then $7 \mid n$.
- By 9: If $9 \mid (a_k + a_{k-1} + \cdots + a_1 + a_0)$, then $9 \mid n$.
- By 11: If $11 \mid (a_0 - a_1 + a_2 - \cdots + (-1)^k a_k)$, then $11 \mid n$.
- We'll prove a few of the rules below, and will revisit some of the rules later in the course.

GCD's and LCM's

- The *greatest common divisor* (GCD) of m and n (denoted $\gcd(m,n)$) is the largest number d such that $d \mid m$ and $d \mid n$.
- We call two numbers m, n *relatively prime* if $\gcd(m,n) = 1$.
- The *least common multiple* (LCM) of m and n (denoted $\operatorname{lcm}(m,n)$) is the smallest number l such that $m \mid l$ and $n \mid l$.

2.1 Example Questions

Problem 2.1 Formally prove the two "easy facts" mentioned earlier. Note: These are not hard to prove, but it is useful to see it formally written out.

(a) If $a \mid n$ and $a \mid m$ then $a \mid n \pm m$.

(b) If $a \mid n$ and k is any integer, then $a \mid k \cdot n$.

Problem 2.2 Prove the divisibility rule for:

(a) 5

(b) 2^j

(c) 9 when $n = \overline{abcde}$. Note: The proof for the general case is the same idea, but harder to write down neatly.

Problem 2.3 Consider the integer $\overline{2a3a1a}$.

(a) If the number is divisible by 9, what are the possible values for a?

(b) If the number is divisible by 11, what are the possible values for a?

Problem 2.4 Prove the divisibility rule for 7. That is, prove that $n = 10a + b$ is divisible by 7 if and only if $a - 2b$ is divisible by 7.

Problem 2.5 Possible Numbers

(a) (AIME 1984) The integer n is the smallest positive multiple of 15 such that every digit of n is either 0 or 8. Find n.

(b) Find all possible numbers between 5000 and 10000 that are divisible by $2, 3, 5, 9, 11$.

Problem 2.6 A five-digit number has five distinct digits, and it is divisible by 9. What is the largest such number?

Problem 2.7 Let k be an even number. Is it possible to write 1 as the sum of the reciprocals of k odd integers?

Problem 2.8 Prove that $\gcd(b, a) = \gcd(b - a, a)$. Hint: For integers m, n if $m \mid n$ and $n \mid m$ then $m = n$.

Problem 2.9 Every number less than 100000 can be written as $\overline{abcde}$ where $a, b, c, d, e \in \{0, 1, \ldots, 9\}$. For how many of these numbers is it true that 11 divides the sum of $\overline{abc}$ and $\overline{de}$.

Problem 2.10 Given a six-digit number $\overline{abcdef}$, whose digits are 1,2,3,4,5,6, not necessarily in this order. Assume that $6 \mid \overline{abcdef}$, $5 \mid \overline{abcde}$, $4 \mid \overline{abcd}$, $3 \mid \overline{abc}$, and $2 \mid \overline{ab}$. Find $\overline{abcdef}$.

2.2 Quick Response Questions

Problem 2.11 Is the number 13531 divisible by 3?

Problem 2.12 Is the number 1598751 divisible by 9?

Problem 2.13 Does 7 divide 51342?

Problem 2.14 Is 3740 divisible by 99?

Problem 2.15 What is the largest 3-digit number divisible by 2, 3, 7, and 9?

Problem 2.16 Find the number closest to 12345 that is divisible by 9. (If 12345 is divisible by 9, then input 12345.)

Problem 2.17 Find the largest power of 2 (1, 2, 4, 8, etc.) that divides 2121212.

Problem 2.18 Find the smallest number made up of the digits $1, 2, 3, 4$ that is divisible by 11?

Problem 2.19 Find the smallest number consisting of only 1's that is divisible by 9 and 11. How many digits does this number have?

Problem 2.20 What is the last digit of 1224 when written in base 3?

2.3 Practice Questions

Problem 2.21 Explain how to get for divisibility by 6. Explain why your method works using the facts mentioned in class (you do not need to prove these facts).

Problem 2.22 Prove the divisibility rule for 11 when $n = \overline{abcd}$.

Problem 2.23 Consider the integer $n = \overline{1a2a3a4a}$.

(a) Find all a such that $9 \mid n$.

(b) Find all a such that $11 \mid n$.

Problem 2.24 Another possible divisibility rule for 7 is: $10a + b$ is divisible by 7 if and only if $3a + b$ is divisible by 7. Prove this rule, and compare it to the rule presented above (are there any advantages or disadvantages to this rule?).

Problem 2.25 Find the largest multiple of 11 that is a four digit number $abcd$ with distinct odd digits.

Problem 2.26 A five-digit number, all of whose digits are distinct, is divisible by 11. Given that its left-most digit is 3. What is the smallest such number?

Problem 2.27 Suppose that n is an odd positive integer and $n = a^2 + b^2$ for some pair of positive integers a, b. Show that $n = 4m + 1$ for some positive integer m.

Problem 2.28 Prove that if $b = aq + r$, then $\gcd(b, a) = \gcd(a, r)$ by

(a) using that $\gcd(b, a) = \gcd(b - a, a)$.

(b) using a similar method as the one used to show that $\gcd(b, a) = \gcd(b - a, a)$ (without using that $\gcd(b, a) = \gcd(b - a, a)$).

Problem 2.29 How many 5-digit numbers $\overline{abcde}$ (so $a, b, c, d, e \in \{0, 1, \ldots, 9\}$ with $a \neq 0$) are there such that the sum of $\overline{abcd}$ and $\overline{bcde}$ is divisible by 11?

Problem 2.30 Let $\overline{a357b}$ be a five-digit number. If $44 \mid \overline{a357b}$, find the values of a and b.

3. Prime Numbers

Prime Factorization

- **Euclid**: There are infinitely many primes. (You do not need to know the proof.)
- **Unique Prime Factorization**: Every positive integer n has a unique factorization as a product of primes. That is,

$$n = p_1^{e_1} p_2^{e_2} \cdots p_k^{e_k}$$

  for distinct primes p_i and $e_i > 0$. (You do not need to know the proof.)
- **Divisors**: If $n = p_1^{e_1} p_2^{e_2} \cdots p_k^{e_k}$ and $a \mid n$, then $a = p_1^{f_1} p_2^{f_2} \cdots p_k^{f_k}$ where $0 \le f_i \le e_i$.
- **Number of factors**: For an integer n, $\sigma_0(n)$ is the number of factors of n. If $n = p_1^{e_1} p_2^{e_2} \cdots p_k^{e_k}$ is the prime factorization of n, then

$$\sigma_0(n) = (e_1 + 1)(e_2 + 1) \cdots (e_k + 1) = \prod_{i=1}^{k}(e_i + 1).$$

GCD's and LCM's

- The *greatest common divisor* (GCD) of m and n (denoted $\gcd(m,n)$) is the largest number d such that $d \mid m$ and $d \mid n$.
- We call two numbers m, n *relatively prime* if $\gcd(m,n) = 1$.
- The *least common multiple* (LCM) of m and n (denoted $\text{lcm}(m,n)$) is the smallest number l such that $m \mid l$ and $n \mid l$.
- If m and n be positive integers with prime factorizations $m = p_1^{e_1} p_2^{e_2} \cdots p_k^{e_k}$ and $n = p_1^{f_1} p_2^{f_2} \cdots p_k^{f_k}$ (here, e_i and f_i can equal zero) then

- $\circ$ $\gcd(m,n) = \prod_{i=1}^{k} p_i^{\min(e_i, f_i)}$.
- $\circ$ $\text{lcm}(m,n) = \prod_{i=1}^{k} p_i^{\max(e_i, f_i)}$.
- The following facts are also true (to be proven in the problems):
  - $\circ$ $m \cdot n = \gcd(m,n) \cdot \text{lcm}(m,n)$.
  - $\circ$ If $a \mid m$ and $a \mid n$, then $a \mid \gcd(m,n)$.
  - $\circ$ If $m \mid k$ and $n \mid k$, then $\text{lcm}(m,n) \mid k$.

Euclidean Algorithm

- Recall the division algorithm: For any integers a and b, $a \neq 0$, there exists a unique pair (q,r) of integers such that $b = aq + r$ and $0 \leq r < |a|$. The number q is called the *quotient*, and r is called the *remainder*.
- Idea behind Euclidean Algorithm: For any integers b, a,
  - $\circ$ $\gcd(b,a) = \gcd(b-a, a)$.
  - $\circ$ If $b = aq + r$ as in the division algorithm, $\gcd(b,a) = \gcd(a,r)$.
- Euclidean Algorithm: Assume $m \geq n$ and let $r_0 = m$ and $r_1 = n$, then there exists sequences of integers $q_1, \ldots, q_k$ and $r_2, \ldots, r_k$ such that:

$$
\begin{aligned}
r_0 &= r_1 q_1 + r_2 & \text{with} \quad & 0 < r_2 < r_1 \\
r_1 &= r_2 q_2 + r_3 & \text{with} \quad & 0 < r_3 < r_2 \\
&\vdots & & \vdots \\
r_{k-1} &= r_k q_k & \text{with} \quad & r_k = \gcd(m,n)
\end{aligned}
$$

- For example, to find the greatest common divisor of 900 and 243 using the Euclidean algorithm:

$$
\begin{aligned}
900 &= 243 \times 3 + 171, \\
243 &= 171 \times 1 + 72, \\
171 &= 72 \times 2 + 27, \\
72 &= 27 \times 2 + 18, \\
27 &= 18 \times 1 + 9, \\
18 &= 9 \times 2. \quad \text{(Remainder is 0, so the last divisor 9 is our answer)}
\end{aligned}
$$

and therefore $\gcd(900, 243) = 9$.

- In other words, $\gcd(900, 243) = \gcd(243, 171) = \gcd(171, 72) = \gcd(72, 27) = \gcd(27, 18) = \gcd(18, 9) = 9$.

Bezout's Identity

- For any two positive integers m and n, there exist integers a and b such that $am + bn = \gcd(m,n)$.
- For example, if $m = 9, n = 11$ so $\gcd(m,n) = 1$, we have pairs such as $(5,-4), (-6,5)$ that work for (a,b).
- In fact there are infinitely many pairs (a,b) for any m,n.
- **Note**: The Euclidean Algorithm can be extended (to the "Extended Euclidean Algorithm") to generate an example of a,b (and therefore proving the theorem as well). This is beyond the scope of this class. Nevertheless, you may use Bezout's Identity when applicable.

3.1 Example Questions

Problem 3.1 Prove the following:

(a) The "number of factors" formula above.

(b) A number has an odd number of factors if and only if it is a square.

Problem 3.2 Find the smallest positive integer n such that

(a) $\sqrt{200n}$ is an integer.

(b) $\sqrt{200n}$ is a perfect cube.

Problem 3.3 Compute the product of all distinct positive divisors of 120^6 (express your answer as a power of 120).

Problem 3.4 Prove the following:

(a) $m \cdot n = \gcd(m,n) \cdot \mathrm{lcm}(m,n)$.

(b) If $a \mid m$ and $a \mid n$, then $a \mid \gcd(m,n)$.

Problem 3.5 Do the following:

(a) Find the greatest common divisor of $2^{2016} - 1$ and $2^{100} - 1$.

(b) Show that the fraction
$$\frac{15n + 4}{3n + 1}$$
is irreducible for all positive integers n.

Problem 3.6 Application of Bezout's Identity

(a) Let $d = \gcd(m,n)$. Prove that if $k = am + bn$ for any integers a, b, then $d \mid k$. Hint: This is not hard, don't overthink!

(b) Suppose a society only has bills of value 34 and 62. Suppose everyone in the society always carries around plenty of both bills. Find all (integer value) prices that it is possible to purchase in this society.

Problem 3.7 Find all numbers n less than 50 with the following property: the product of the divisors of n is equal to n^2.

Problem 3.8 Let x, y be positive integers, $x < y$, and $x + y = 667$. Given that $\dfrac{\text{lcm}(x,y)}{\gcd(x,y)} = 120$. Find all such pairs (x, y).

Problem 3.9 From the set $\{1, 2, \ldots, 100\}$, select k numbers. What is the minimum value of k such that it is guaranteed to have two numbers that are not relatively prime? Hint: How many prime numbers are there less than 100?

Problem 3.10 (Putnam 2000) Given integers n, m, $n \geq m \geq 1$. Show that $\dfrac{\gcd(m,n)}{n} \dbinom{n}{m}$ is an integer. Hint: Recall $\dbinom{n}{m} = \dfrac{n!}{m!(n-m)!}$.

3.2 Quick Response Questions

Problem 3.11 Find the prime factorization of 1683. What is the largest prime appearing in the prime factorization?

Problem 3.12 Find the prime factorization of 7425. How many distinct primes appear in the factorization?

Problem 3.13 Find the number of factors of 1683.

Problem 3.14 Find the number of factors 13005.

Problem 3.15 Find $\gcd(2310, 13005)$.

Problem 3.16 Find $\operatorname{lcm}(2310, 13005)$.

Problem 3.17 Find $\gcd(4200, 3430)$.

Problem 3.18 Find $\operatorname{lcm}(4200, 3430)$.

Problem 3.19 The product of all the factors of 2^9 is 2^M for M in integer. What is M?

Problem 3.20 What is the smallest number with 10 factors?

3.3 Practice Questions

Problem 3.21 Define the three term GCD function $\gcd(a,b,c)$ as the largest common divisor of a,b,c. Prove that $\gcd(a,b,c) = \gcd(\gcd(a,b),c)$. Hint: It may be useful to first prove a formula for $\gcd(a,b,c)$ using prime factorizations.

Problem 3.22 Find the smallest perfect square that is a multiple of both 360 and 525.

Problem 3.23 Compute the product of all distinct positive divisors that are perfect cubes of 120^6. Express your answer as a power of 120.

Problem 3.24 Prove that if $m \mid k$ and $n \mid k$, then $\text{lcm}(m,n) \mid k$.

Problem 3.25 Find the greatest common divisor of 20162017 and 20152016.

Problem 3.26 Suppose a society only has three dollar and eleven dollar bills. George has 500 three dollar bills and 100 eleven dollar bills when he goes shopping at a department store, where he will spend up to 100 dollars.

(a) Prove that if the department store can give change (still using three dollar and eleven dollar bills), it is possible for George to buy something from the store that costs and dollar amount from \$1 to \$100. As examples, show how George can purchase something that costs \$40 and something that costs \$19.

(b) If the department store cannot give change, there are dollar amounts that George cannot pay for. What is the largest price D that George cannot pay? (Still assume $1 \le D \le 100$ with D an integer.)

Problem 3.27 Find all numbers n less than 50 with the following property: the product of the divisors is equal to n^3.

Problem 3.28 Find all ordered pairs of positive integers (x, y) such that $1! + 2! + 3! + \cdots + x! = y^2$.

Problem 3.29 From the set $\{1, 2, \ldots, 100\}$, select k numbers. What is the minimum value of k such that it is guaranteed to have two numbers with 3 as a common divisor?

Problem 3.30 Use the following outline to prove Bezout's Identity.

(a) There are positive integers of the form $j = am + bn$.

(b) If there are positive integers of the form $j = am + bn$ there is a smallest positive integer e of the form $e = a'm + b'n$ for integers a', b'.

(c) Use the division algorithm to show that e divides both m and n.

(d) Show that $\gcd(m, n)$ must divide e, hence $e = \gcd(m, n)$.

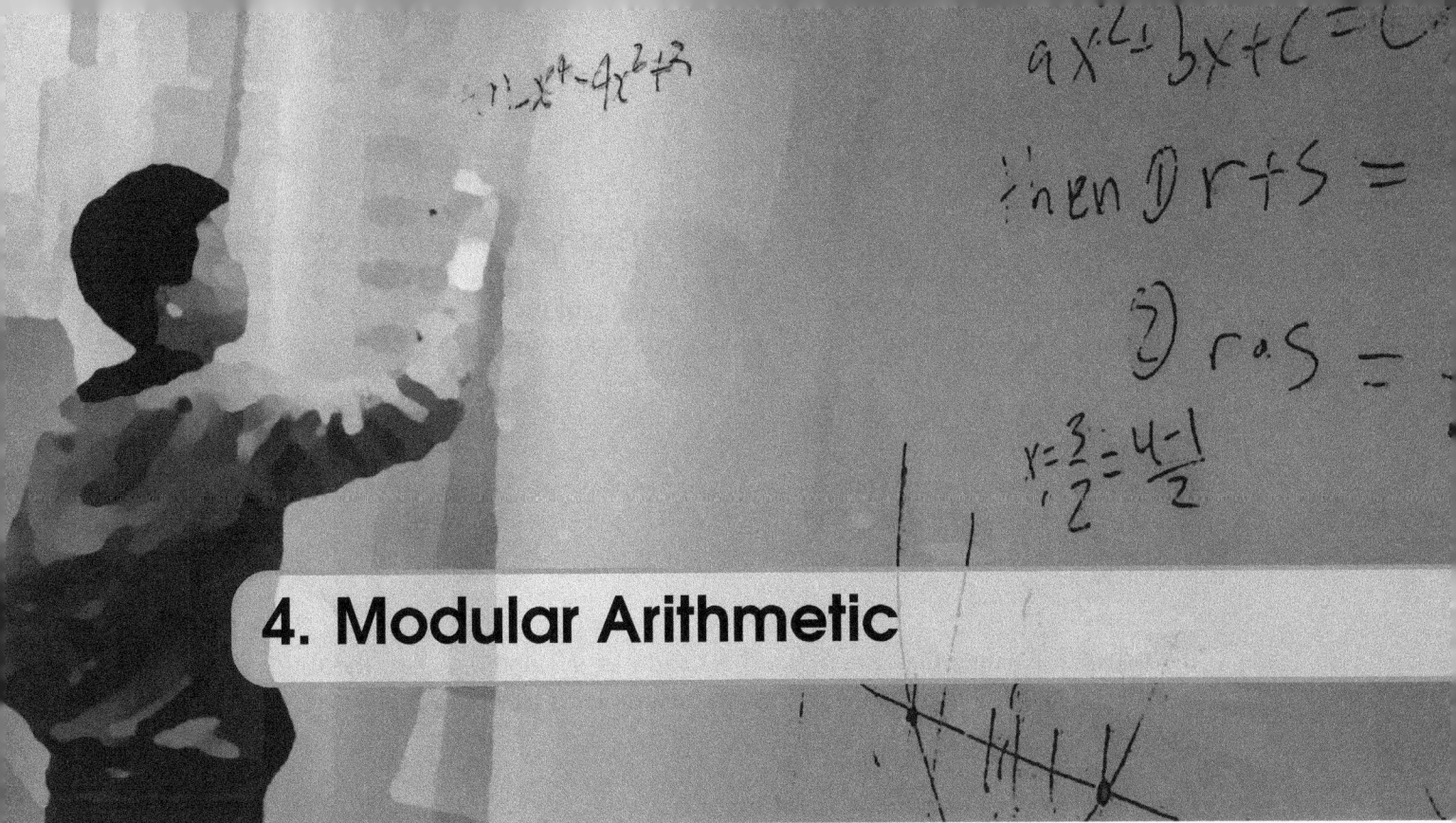

4. Modular Arithmetic

Modular Arithmetic

- Two numbers a and b are *congruent modulo* m (denoted $a \equiv b \pmod{m}$) if $m \mid (a - b)$.
- Equivalently, a and b have the same remainder when divided by m.
- If we are working modulo m, we often call m the *modulus*.
- In other words, a and b have the same remainder when divided by m.
- For example, there are 7 days per week; use 0 for Sunday, 1 for Monday, ..., 6 for Saturday. Thus, days of the week can be thought of using a modulus of 7. Suppose today is March 14, a Saturday. March 19 is 5 days from now, so it will be $6 + 5 = 11$, and $11 \equiv 4 \pmod{7}$, thus it is Thursday.
- The "Clock Arithmetic" is also an example of modular arithmetic. The modulus is 12. Suppose it is 11 o'clock now. 6 hours from now, $11 + 6 = 17 \equiv 5 \pmod{12}$, so it will be 5 o'clock (switched am/pm). If we want to calculate based on 24-hour time, then use modulo 24.

4.1 Example Questions

Problem 4.1 Warmups

(a) Prove the equivalence mentioned in the beginning of the packet: $m \mid (a - b)$ if and only if a and b have the same remainder when divided by m.

(b) Prove that any year (including a leap year) must have at least one "Friday the 13th".

Problem 4.2 Assume that $a \equiv b \pmod{m}$ and $c \equiv d \pmod{m}$. Are the following true or false? If false, come up with a counterexample. If true, you'll prove it on your homework!

(a) If $b \equiv c \pmod{m}$ then $a \equiv c \pmod{m}$.

(b) $(a + c) \equiv (b + d) \pmod{m}$.

(c) $(a \cdot c) \equiv (b \cdot d) \pmod{m}$.

(d) If k is an integer and $k \mid a, k \mid b$, then $(a/k) \equiv (b/k) \pmod{m}$.

(e) If n is a positive integer, then $a^n \equiv b^n$.

Problem 4.3 Calculations

(a) What is the units digit of 2^{2016}?

(b) Find the remainder when $31^{999} + 65^{100}$ is divided by 32.

(c) What is the units digit of $1^2 + 2^2 + 3^2 + \cdots + 99^2$?

Problem 4.4 Answer the following.

(a) A certain natural number n has a unit digit 9 when expressed in base 12. Find the remainder when n^2 is divided by 6.

(b) If $m > 1$ and $69 \equiv 90 \equiv 125 \pmod{m}$, what is m?

Problem 4.5 Let $n = \overline{a_k a_{k-1} \ldots a_1 a_0} = a_k 10^k + a_{k-1} 10^{k-1} + \cdots + a_1 10 + a_0, a_i \in \{0, 1, \ldots, 9\}$. Prove the following. Note: These are similar to things you've already proven, but practice using modular arithmetic here!

(a) Prove that $n \equiv \overline{a_{j-1} a_{j-2} \ldots a_1 a_0} \pmod{2^j}$.

(b) Prove that $n \equiv (a_k + a_{k-1} + \cdots + a_1 + a_0) \pmod 9$. Note this means that a number is equal to the sum of its digits modulo 9.

Problem 4.6 Suppose you create a 15 digit number using five 1's, five 2's, and five 3's. Is it possible that your number is a perfect square?

Problem 4.7 The number 2^{29} is a nine-digit number all of whose digits are distinct. Without computing the actual number, determine which of the ten digits is missing.

Problem 4.8 The Fibonacci sequence is defined by $F_1 = F_2 = 1$, and $F_{n+2} = F_{n+1} + F_n$, that is, the first two terms are both 1, and each subsequence term is the sum of the previous two terms. Find the remainder when F_{2016} is divided by 7.

Problem 4.9 What is the smallest five-digit integer divisible by both 8 and by 9?

Problem 4.10 There are two two-digit numbers whose square ends in the same two-digit number. Find them.

4.2 Quick Response Questions

Problem 4.11 A certain store is running a promotion that starts at 12 AM on Friday. If the promotion runs for 80 hours, what hour will it be when the promotion ends? (Ignore AM or PM in your answer.)

Problem 4.12 If January 1st is a Monday, what day of the week is February 1st (of the same year)? Use Sunday $= 0$, Monday $= 1$, etc. to input your answer.

Problem 4.13 Are 345 and 543 congruent modulo 13?

Problem 4.14 Are 41 and -56 congruent modulo 97?

Problem 4.15 Find the remainder when 51374948 is divided by 9.

Problem 4.16 What is the units digit of 2^{827}?

Problem 4.17 Find all solutions to $n^2 \equiv n \pmod{10}$ with $0 \leq n \leq 9$. How many solutions are there?

Problem 4.18 What is the remainder of 576^2 upon division by 10?

Problem 4.19 What is the remainder of $1! + 2! + 3! + \cdots + 2016!$ upon division by 5?

Problem 4.20 What is $2^{12} - 1 \pmod{13}$.

4.3 Practice Questions

Problem 4.21 What day of the week does June 1st of 2016 fall on?

Problem 4.22 Assume that $a \equiv b \pmod{m}$ and $c \equiv d \pmod{m}$. Prove that

(a) If $b \equiv c \pmod{m}$ then $a \equiv c \pmod{m}$.

(b) $(a+c) \equiv (b+d) \pmod{m}$.

(c) $(a \cdot c) \equiv (b \cdot d) \pmod{m}$.

(d) If n is a positive integer, then $a^n \equiv b^n$.

Problem 4.23 Calculations Continued

(a) What is the units digit of $3^{1234} + 7^{4321}$?

(b) Is $43^{101} + 23^{101}$ divisible by 66?

(c) What is the remainder when $1^2 + 2^2 + \cdots + 99^2$ is divided by 7?

Problem 4.24 Find a digit d ($0 \le d \le 9$) so that $171 \times 541 = 92d11$. Try to do this without multiplying out the numbers!

Problem 4.25 Let $n = \overline{a_k a_{k-1} \ldots a_1 a_0} = a_k 10^k + a_{k-1} 10^{k-1} + \cdots + a_1 10 + a_0, a_i \in \{0, 1, \ldots, 9\}$. Prove the following. Note: These are similar to things you've already proven, but practice using modular arithmetic here!

(a) Prove that $n \equiv \overline{a_k + a_{k-1} + \cdots + a_1 + a_0}$ (mod 3).

(b) Prove that $n \equiv (a_0 - a_1 + a_2 \cdots + (-1)^k a_k)$ (mod 11).

Problem 4.26 Find the remainder when the sum $52 + 522 + 5222 + \cdots + 522\cdots22$ (the last term has twenty 2's) is divided by 11.

Problem 4.27 Find the remainder of 7^{101} upon division by 19.

Problem 4.28 Consider the sequence a_n defined by $a_1 = a_2 = 1$ and $a_n = 2a_{n-1} + a_{n-2}$. Find the remainder when a_{2015} is divided by 4.

Problem 4.29 A five digit number $\overline{4a77b}$ is divisible by 99, find the values of a, b.

Problem 4.30 A two digit number equals the sum of its tens digit and the square of its units digit. What is this two digit number?

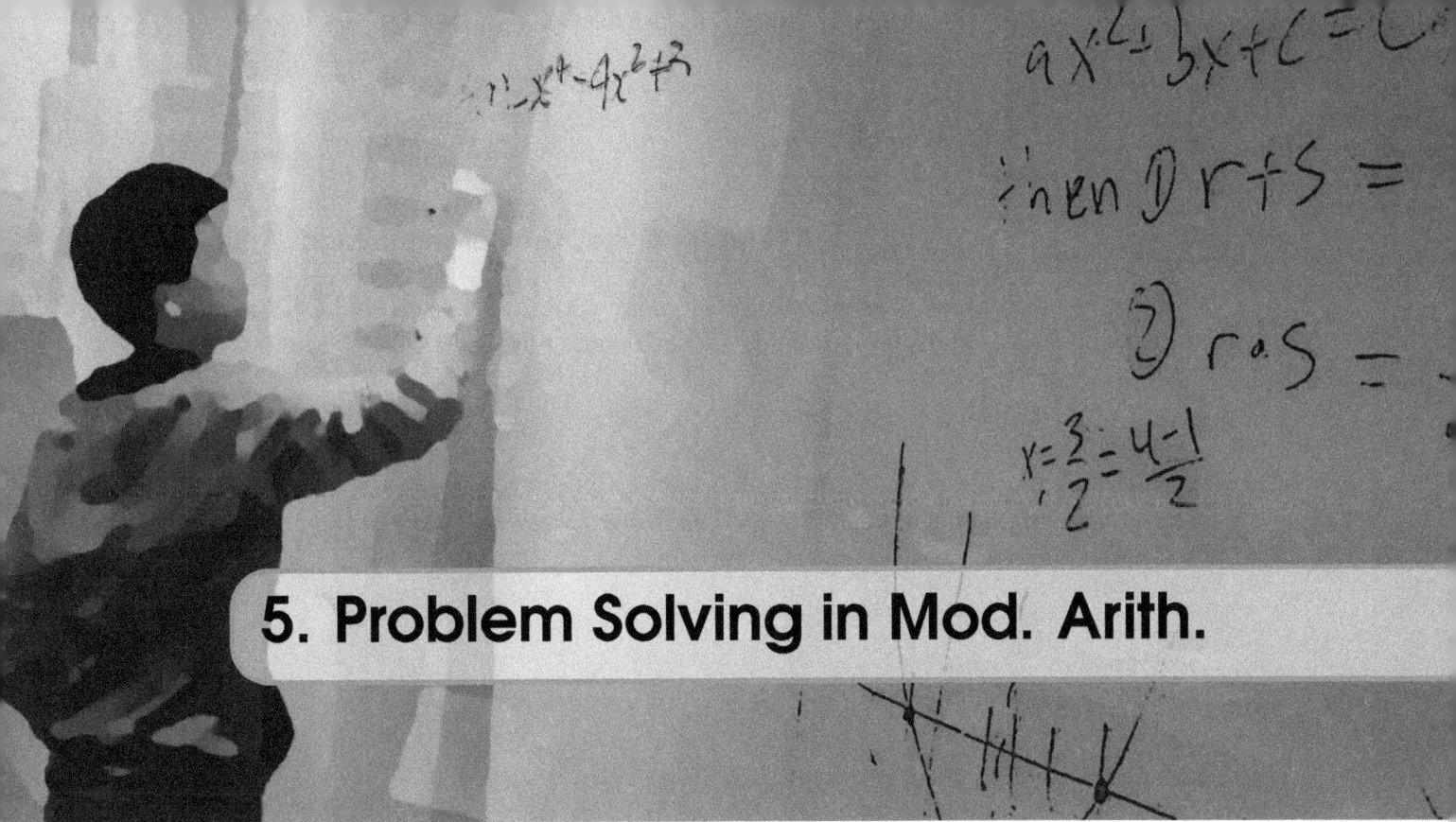

5. Problem Solving in Mod. Arith.

Modular Arithmetic Review

- Two numbers a and b are congruent modulo m (denoted $a \equiv b \pmod{m}$) if $m \mid (a-b)$; in other words, a and b have the same remainder when divided by m.
- Equality, addition, and multiplication all "work" (more formally are well-defined):
  - If $a \equiv b \pmod{m}$ and $b \equiv c \pmod{m}$, then $a \equiv c \pmod{m}$.
  - If $a \equiv b \pmod{m}$ and $c \equiv d \pmod{m}$, then $(a \pm c) \equiv (b \pm d) \pmod{m}$.
  - If $a \equiv b \pmod{m}$ and $c \equiv d \pmod{m}$, then $ac \equiv bd \pmod{m}$.
- We saw last week we needed to be a little more careful with division. We'll prove the following facts in this handout:
  - If $a \equiv b \pmod{m}$ and $k \mid m$, then $a \equiv b \pmod{k}$.
  - If $a \equiv b \pmod{m}$, $d \mid a$, $d \mid b$, and $\gcd(d,m) = 1$, then $\dfrac{a}{d} \equiv \dfrac{b}{d} \pmod{m}$.
  - If $a \equiv b \pmod{m}$, $d \mid a$, and $d \mid b$, then $\dfrac{a}{d} \equiv \dfrac{b}{d} \pmod{\dfrac{m}{\gcd(d,m)}}$.

5.1 Example Questions

Problem 5.1 Modular Multiplicative Inverse

(a) Suppose $\gcd(a,m) = 1$. Consider the set of values $\{0, a, 2a, 3a, \ldots, (m-1)a\}$. Prove that all (m) values have distinct remainders when dividing by m.

(b) Still assume $\gcd(a,m) = 1$. Prove that there is an integer b with $0 \le b < m$ and $ba \equiv 1 \pmod{m}$. Such a b is called the *modular multiplicative inverse* of a modulo m.

Problem 5.2 Prove the following.

(a) If $a \equiv b \pmod{m}$ and $k \mid m$, then $a \equiv b \pmod{k}$.

(b) If $a \equiv b \pmod{m}$, $d \mid a$, $d \mid b$, and $\gcd(d,m) = 1$, then $\dfrac{a}{d} \equiv \dfrac{b}{d} \pmod{m}$.

Problem 5.3 Modular Multiplicative Inverse Practice

(a) Find the modular multiplicative inverse of 8 mod 11.

(b) Find the modular multiplicative inverse of 8 mod 35.

(c) Find the modular multiplicative inverse of 9 mod 70.

(d) Find all solutions to $7x \equiv 3 \pmod{11}$.

Problem 5.4 Find the possible remainders of n^2 in

(a) (mod 3).

(b) (mod 5).

(c) (mod 9).

Problem 5.5 Let $j \geq 2$ be an integer. If n is an integer, find the possible remainders of n^j in (mod 4).

Problem 5.6 Let $k \neq 2$ be the product of the first several primes (that is $k = 2 \cdot 3, k = 2 \cdot 3 \cdot 5$, etc.). Prove that *none* of $k, k-1, k+1$ are perfect squares.

Problem 5.7 Sum of Squares

(a) Is it possible to find two integers n and m such that $n^2 + m^2 = 2015$?

(b) Prove that there are infinitely many numbers that are not the sum of two squares.

Problem 5.8 Can a 5-digit number consisting only of distinct even digits be a perfect square?

Problem 5.9 Do the following equations have integer solutions?

(a) $x^2 - 5y = 102$.

(b) $x^2 - 5y = 104$.

Problem 5.10 Consider the sum of the digits of a perfect square. Is it possible for the sum of the digits to be 2015?

5.2 Quick Response Questions

Problem 5.11 Find the number of factors of 9000.

Problem 5.12 Find the gcd of 140 and 180 using the Euclidean algorithm.

Problem 5.13 What is the modular multiplicative inverse of 5 modulo 12?

Problem 5.14 What is the modular multiplicative inverse of 7 modulo 60?

Problem 5.15 Find the smallest integer $N \geq 500$ such that N^2 leaves remainder 1 when divided by 3.

Problem 5.16 Find the smallest integer $N \geq 500$ such that N^2 leaves remainder 1 when divided by 9.

Problem 5.17 Let p be a prime and m, n be integers. Suppose further that $p^2 \mid m \cdot n$. Is it true that $p \mid m$ or $p \mid n$?

Problem 5.18 Let p be a prime and m, n be integers. Suppose further that $p^2 \mid m \cdot n$ and m, n are relatively prime. Is it true that $p \mid m$ or $p \mid n$?

Problem 5.19 $70 \equiv 10 \pmod{12}$. Let m be a factor of 12. What is the largest m such that $70 \div 5 \equiv 10 \div 5 \pmod{m}$?

Problem 5.20 $70 \equiv 10 \pmod{12}$. Let m be a factor of 12. What is the largest m such that $70 \div 2 \equiv 10 \div 2 \pmod{m}$?

5.3 Practice Questions

Problem 5.21 Inverse Practice

(a) Find $n \cdot a \pmod{12}$ for $0 \le n < 12$ and $a = 5, 8$.

(b) Find all n with $0 \le n < 12$ relatively prime to 12 and their modular multiplicative inverses.

Problem 5.22 Prove that if $a \equiv b \pmod{m}$, $d \mid a$, and $d \mid b$, then $\dfrac{a}{d} \equiv \dfrac{b}{d}$ $\left(\bmod \dfrac{m}{\gcd(d,m)}\right)$.

Problem 5.23

(a) Find all a with $0 \le a < 15$ such that $\gcd(a, 15) = 1$.

(b) Find the modular multiplicative inverses of the numbers in Part (a).

Problem 5.24 Find all possible values of $n^3 \pmod 9$.

Problem 5.25 Show that $(a+b)^3 \equiv a^3 + b^3 \pmod 3$ and $(a+b)^5 \equiv a^5 + b^5 \pmod 5$ for all integers a, b.

Problem 5.26 Make a chart containing the possible remainders of n^2 (mod m) for $m = 2, 3, \ldots, 9$.

Problem 5.27 Prove that there are infinitely many numbers that are not the sum of three squares.

Problem 5.28 Can a 5-digit number consisting only of distinct odd digits be a perfect cube?

Problem 5.29 Does the equation $x^2 - 2y^2 = 1$ have integer solutions?

Problem 5.30 Is it possible for the sum of the digits of a perfect cube to be 2014?

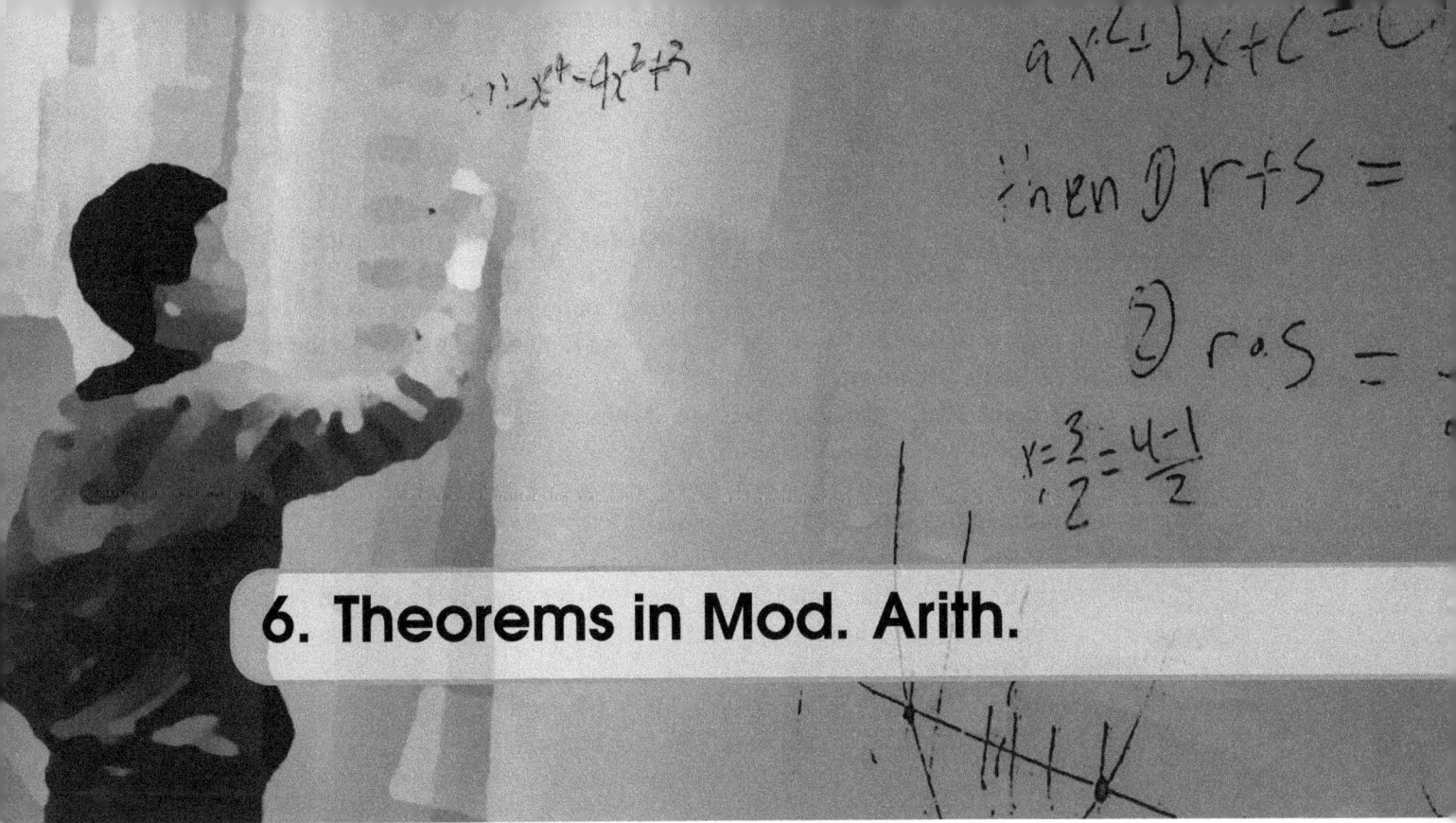

6. Theorems in Mod. Arith.

Residue Classes

- A set S of integers is called a *complete set of residue classes modulo m* if for each $0 \leq i \leq m-1$, there is an element $s \in S$ such that $i \equiv s \pmod{m}$. Also called a *complete residue system modulo m*.
- For example:
  - For any integer a, $\{a, a+1, a+2, \ldots, a+m-1\}$ is a complete set of residue classes modulo m.
  - In particular, we call $\{0, 1, \ldots, m-1\}$ the *minimal nonnegative complete set of residue classes* modulo m.
  - It is common to consider the complete set of residue classes $\{0, \pm 1, \pm 2, \ldots, \pm k\}$ for $m = 2k+1$ and $\{0, \pm 1, \pm 2, \ldots, \pm(k-1), k\}$ for $m = 2k$.

Multiplicative Order of a Number

- Suppose $\gcd(a, m) = 1$. The *(multiplicative) order* of a modulo m is the smallest $k > 0$ such that $a^k \equiv 1 \pmod{m}$.
- For example, if $m = 7$, the powers of 2 mod 7 are $2, 4, 1$, so 2 has order 3 modulo 7.
- Notice the similarity to finding patterns to help calculate large powers (for example, what is the units digit of 3^{100}).

Two Important Theorems

- It is required to understand the proofs of these theorems (given as exercises below), and know how the theorems are applied. It is not required to memorize the proofs.
- **Fermat's Little Theorem**
  - Theorem: If p is prime and a is any integer, then $p \mid (a^p - a)$. Equivalently, if p does not divide a, then $a^{p-1} \equiv 1 \pmod{p}$.
  - For example, since 37 is prime, we can use Fermat's Little Theorem to immediately know that $2^{36} \equiv 1 \pmod{37}$.
  - Note that $p - 1$ is not necessarily the order of a modulo p. For example, Fermat's Little Theorem tells us that $2^{30} \equiv 1 \pmod{31}$ (which is true!). However, $2^5 \equiv 1 \pmod{31}$ as well.
- **Wilson's Theorem**
  - Theorem: For every prime p, $(p-1)! \equiv -1 \pmod{p}$.
  - For example, if $p = 7$ then $6! \equiv 720 \equiv -1 \pmod{7}$.

6.1 Example Questions

Problem 6.1 Order Sanity Check

(a) Suppose a is an integer and we are working mod m. Prove that the sequence $a, a^2, a^3, a^4, \ldots$ eventually starts repeating.

(b) Suppose $\gcd(a, m) = 1$. Show that there is some K (and hence a smallest K) such that $a^K = 1$.

Problem 6.2 Suppose $\gcd(a, m) = 1$. Consider the set $\mathscr{S} = \{a, a^2, a^3, a^4, \ldots\}$.

(a) Fill in the blank to get a true statement: The set S is a complete set of nonzero residue classes modulo m if and only if the multiplicative order of a modulo m is __.

(b) Give an example of a prime m so that the set $\{2, 2^2, \ldots\}$ is a complete set of nonzero residue classes modulo m.

(c) Give an example of a prime m so that the set $\{2, 2^2, \ldots\}$ is *not* a complete set of nonzero residue classes modulo m.

Problem 6.3 Prove Fermat's Little Theorem (if a prime p does not divide a, then $a^{p-1} \equiv 1 \pmod{p}$). Hint: Recall a previous result involving the set $\{a, 2a, 3a, \ldots, (p-1)a\}$.

Problem 6.4 Wilson's Theorem

(a) Suppose p is a prime. Show that $x^2 \equiv 1 \pmod{p}$ has exactly 2 solutions modulo p. Hint: Consider the minimal nonnegative complete set of residue classes modulo p.

(b) Prove Wilson's Theorem (for a prime p, $(p-1)! \equiv -1 \pmod{p}$). Hint: Recall multiplicative inverses.

Problem 6.5 Compute the following.

(a) The remainder when 9^{2016} is divided by 11.

(b) The remainder of $2^{50} + 3^{50}$ when divided by 13.

Problem 6.6 One Application of Wilson's Theorem

(a) Find the modular multiplicative inverse of 2 mod 67.

(b) Calculate 64! (mod 67).

Problem 6.7 Suppose a sequence has recursive definition $a_1 = k$ (k a positive integer) and $a_{n+1} = 5^{a_n}$. Find the remainder when a_{100} is divided by 11.

Problem 6.8 Find all primes p such that $p \mid 3^p + 1$.

Problem 6.9 Assume $p = 4k + 1$ is a prime. Prove that $\left[\left(\dfrac{p-1}{2}\right)!\right]^2 \equiv -1 \pmod{p}$.

Hint: Use Wilson's Theorem and try pairing the number r with $p - r$ in $(p-1)!$.

Problem 6.10 Show that there are infinitely many n such that $3 \mid n2^n + 1$.

6.2 Quick Response Questions

Problem 6.11 What is the units digit of 7^{77}?

Problem 6.12 What is the multiplicative order of 3 modulo 13?

Problem 6.13 What is the multiplicative order of 10 modulo 13?

Problem 6.14 Is there an $k > 0$ such that $3^k \equiv 1 \pmod{10}$?

Problem 6.15 Is there an $k > 0$ such that $4^k \equiv 1 \pmod{10}$?

Problem 6.16 What is 3^{14} modulo 13?

Problem 6.17 Find $1946^{32} \pmod{31}$.

Problem 6.18 Find the remainder when 33^{1946} is divided by 31.

Problem 6.19 Use Wilson's Theorem to find the remainder of 9! upon division by 11.

Problem 6.20 Calculate $(50!)^2 \pmod{101}$.

6.3 Practice Questions

Problem 6.21 Prove that if $\gcd(a,m) > 1$, then $a^k \not\equiv 1 \pmod{m}$ for all $k \geq 1$. Hint: Consider a proof by contradiction.

Problem 6.22 Suppose $\gcd(a,m) = 1$. Let j be the order of a modulo m. Prove that if $a^k \equiv 1 \pmod{m}$, then $j \mid k$.

Problem 6.23 Prove the following facts about the different forms of Fermat's Little Theorem. Assume p is a prime throughout. Note: All the results are not hard, but are good to write out at least once.

(a) If $a^{p-1} \equiv 1 \pmod{p}$ then $p \mid a^p - a$.

(b) If $p \mid a^p - a$ and p does not divide a then $a^{p-1} \equiv 1 \pmod{p}$.

(c) If p does divide a, then $p \mid a^p - a$ but $a^{p-1} \not\equiv 1 \pmod{p}$.

Problem 6.24 Verify directly that $p \mid (p-1)! + 1$ for $p = 2, 3, 5, 7, 11$.

Problem 6.25 Find the remainder of 2016^{2017} upon division by 17.

Problem 6.26 Wilson's Theorem Practice

(a) Find the modular multiplicative inverse of 17 mod 23.

(b) Calculate 19! (mod 23).

Problem 6.27 Suppose a sequence has recursive definition $a_1 = 2$ and $a_{n+1} = 4^{a_n}$ (mod 11). Find a_{100}.

Problem 6.28 Find all primes p such that $p \mid 2^{2p} + 1$.

Problem 6.29 Assume $p = 4k + 1$ is a prime. It was proven that $\left[\left(\dfrac{p-1}{2} \right)! \right]^2 \equiv -1$ (mod p).

Show that both the fact that p is prime and that $p = 4k + 1$ are important:

(a) Find a number of the form $n = 4k + 1$ (that is not prime) so that $\left[\left(\dfrac{n-1}{2} \right)! \right]^2 \not\equiv -1$ (mod n).

(b) Find a prime number $p > 2$ (not of the form $p = 4k + 1$) so that $\left[\left(\dfrac{p-1}{2} \right)! \right]^2 \not\equiv -1$ (mod p).

Problem 6.30 (Putnam 1972) Show that given an odd prime p, there are infinitely many integers such that $p \mid n2^n + 1$. Hint: $(-1) \cdot 1 + 1 = 0$.

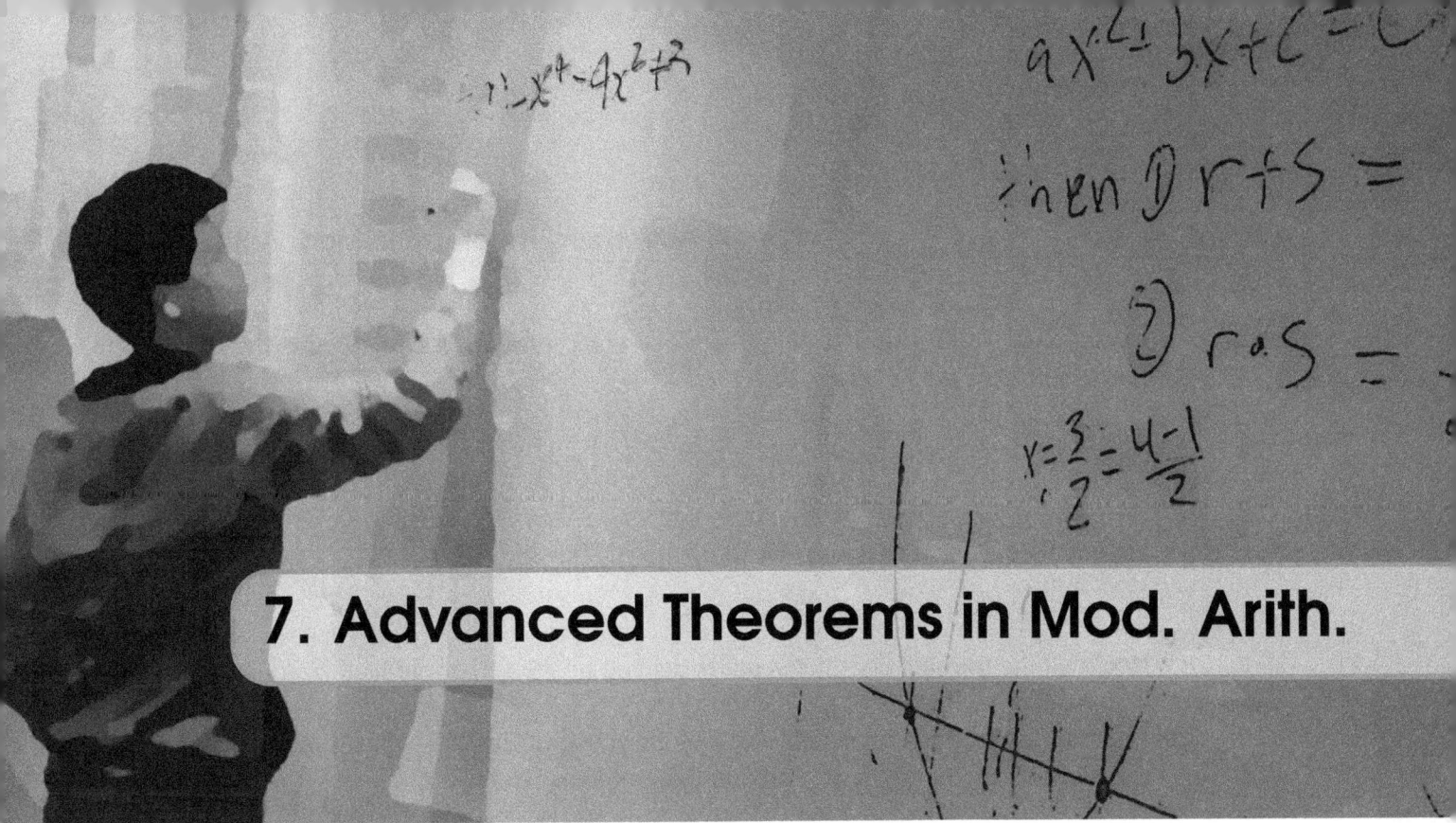

7. Advanced Theorems in Mod. Arith.

Residue Classes

- Recall a set S of integers is called a *complete set of residue classes modulo m* if for each $0 \leq i \leq m-1$, there is an element $s \in S$ such that $i \equiv s \pmod{m}$. Also called a *complete residue system modulo m*.
- We have all of the following sets are complete sets of residue classes modulo m
  - $\{a, a+1, a+2, \ldots, a+m-1\}$ for any integer a (recall if $a = 0$ we call the set the *minimal nonnegative complete set of residue classes*).
  - $\{a, 2a, 3a, \ldots, (m-1)a\}$ for integers a with $\gcd(a,m) = 1$.
  - $\{0, \pm 1, \pm 2, \ldots, \pm k\}$ or $\{0, \pm 1, \pm 2, \ldots, \pm(k-1), k\}$ for $m = 2k+1, 2k$ respectively.
- A set S of integers is called a *reduced set of residue classes modulo m* if for each $0 \leq i \leq m-1$ where $\gcd(i, m) = 1$, there is an element $s \in S$ such that $i \equiv s \pmod{m}$. Also called a *reduced residue system modulo m*.
- For example:
  - $\{1, 3, 7, 9\}$ is a reduced set of residue classes modulo 10.
  - If p is prime, $\{1, 2, \ldots, p-1\}$ is a reduced set of residue classes modulo p.

Euler Totient Function

- The number of elements in a reduced set of residue classes modulo m is denoted as $\phi(m)$. This is known as the *Euler ϕ function*, also called the *Euler totient function*.
- For example, from above we have:
  - $\phi(10) = 4$.

- o If p is prime, then $\phi(p) = p - 1$.
- In fact, there is a formula (called *Euler's product formula*) for calculating the ϕ function. If $n = p_1^{e_1} p_2^{e_2} \cdots p_k^{e_k}$, then

$$\phi(n) = n \left(1 - \frac{1}{p_1}\right) \left(1 - \frac{1}{p_2}\right) \cdots \left(1 - \frac{1}{p_k}\right).$$

- For example, double checking our results from above, $\phi(10) = 10 \cdot \frac{1}{2} \cdot \frac{4}{5} = 4$ and if p is a prime, $\phi(p) = p \cdot \frac{p-1}{p} = p - 1$.
- Note that the formula does not actually depend on the full prime factorization of n, just the prime factors of n.
- The full proof of this formula is beyond the scope of this class, but some of the ideas behind the proof are explored in the problems below.

Extension of Fermat's Little Theorem

- (Euler's Extension to Fermat's Little Theorem.) Let $\gcd(a, n) = 1$. Then $a^{\phi(n)} \equiv 1$ (mod n).
- This extends Fermat's Little Theorem because $\phi(p) = p - 1$ for primes p.
- For example, $\phi(10) = 4$, so $7^4 \equiv 1$ (mod 10). Hence, the units digit of 7^{801} is 7.
- We will not focus on the proof of this theorem in this class. However, the proof follows the same ideas as the proof of Fermat's Little Theorem.

Chinese Remainder Theorem

- Theorem: Let $m_1, \ldots, m_k$ be pairwise relatively prime positive integers (that is, $\gcd(m_i, m_j) = 1$ for all $i \neq j$). Let $b_1, \ldots, b_k$ be arbitrary integers. Then the system

$$
\begin{aligned}
x &\equiv b_1 && (\text{mod } m_1) \\
x &\equiv b_2 && (\text{mod } m_2) \\
&\vdots \quad \vdots \quad \vdots \\
x &\equiv b_k && (\text{mod } m_k)
\end{aligned}
$$

has a unique solution modulo $m_1 m_2 \cdots m_k$.
- For example, the Chinese Remainder Theorem guarantees the existence of an x such that

$$
\begin{aligned}
x &\equiv 2 && (\text{mod } 3) \\
x &\equiv 3 && (\text{mod } 5) \\
x &\equiv 2 && (\text{mod } 7).
\end{aligned}
$$

Furthermore, this x is unique modulo $3 \cdot 5 \cdot 7 = 105$. (In fact, $x = 23$ is a solution.)

7.1 Example Questions

Problem 7.1 Reduced Residue System and Euler Totient Function Practice

(a) Calculate $\phi(8)$ using the formula above and double check your answer by giving an example of a reduced set of residue classes modulo 8.

(b) Calculate $\phi(15)$ using the formula above and double check your answer by giving an example of a reduced set of residue classes modulo 15.

(c) Calculate $\phi(2001)$.

Problem 7.2 If $\gcd(m,n) = 1$, then $\phi(m \cdot n) = \phi(m) \cdot \phi(n)$ (such a function is called *multiplicative*). The proof of this result is beyond the scope of this course, but we'll examine the fact a little below.

(a) Show (by calculating each separately) that $\phi(30) = \phi(5) \cdot \phi(6)$. (You may use the formula for ϕ.)

(b) Show (by calculating each separately) that $\phi(200) = \phi(8) \cdot \phi(25)$. (You may use the formula for ϕ.)

(c) Show that $\phi(200) \neq \phi(20) \cdot \phi(10)$. (You may use the formula for ϕ.)

(d) Prove that $\phi(p \cdot q) = \phi(p) \cdot \phi(q)$ for primes p, q. Do this directly from the definition of ϕ, not using the formula. Hint: There are pq total numbers $\{1, 2, \ldots, pq\}$. How many of them are not in a reduced residue system?

Problem 7.3 Find the last two digits of:

(a) 13^{1998}. Hint: What is the modular multiplicative inverse of 13 modulo 100?

(b) 32^{1998}.

Problem 7.4 Find the remainder of $13^{3^{13}}$ when it is divided by 17.

Problem 7.5 Solve the system

$$\begin{aligned} x &\equiv 1 \pmod{7} \\ x &\equiv 5 \pmod{16} \end{aligned}$$

Problem 7.6 Consider a general system of three equations:

$$\begin{aligned} x &\equiv b_1 \pmod{m_1} \\ x &\equiv b_2 \pmod{m_2} \\ x &\equiv b_3 \pmod{m_3} \end{aligned}$$

with m_1, m_2, m_3 pairwise relatively prime.

(a) Prove that $m_2 \cdot m_3$ has a modular multiplicative inverse modulo m_1. (Similarly, $m_1 \cdot m_3$ has a modular multiplicative inverse modulo m_2 and $m_1 \cdot m_2$ has a modular multiplicative inverse modulo m_3.)

(b) Let c_1 be the modular multiplicative inverse of $m_2 m_3$ modulo m_1 (this exists by part (a)), and similarly let c_2 be the modular multiplicative inverse of $m_1 m_3$ modulo m_2 and let c_3 be the modular multiplicative inverse of $m_1 m_2$ modulo m_3. Show that

$$x = b_1 \cdot c_1 \cdot m_2 m_3 + b_2 \cdot c_2 \cdot m_1 m_3 + b_3 \cdot c_3 \cdot m_1 m_2$$

is a solution to the system of equations.

Problem 7.7 Solve the system

$$
\begin{aligned}
x &\equiv 2 \quad (\text{mod } 5) \\
x &\equiv 4 \quad (\text{mod } 9) \\
x &\equiv 6 \quad (\text{mod } 11).
\end{aligned}
$$

Problem 7.8 Consider the question: "Find the smallest number that leaves a remainder of 1 when divided by 2, 2 when divided by 3, 3 when divided by 4, 4 when divided by 5, and 5 when divided by 6."

(a) Does the Chinese Remainder Theorem apply to the question?

(b) Solve the question.

Problem 7.9 Find the remainder when $7^3 + 7^{7^2} + 7^{7^3} + 7^{3^4} + \cdots + 7^{7^{20}}$ is divided by 11.

Problem 7.10 Let $p \neq q$ are primes with $\gcd(p-1, q-1) = m$. Set $k = (p-1)(q-1)/m$. Prove that if $\gcd(a, pq) = 1$, then $a^k \equiv 1 \pmod{pq}$.

7.2 Quick Response Questions

Problem 7.11 Calculate $\phi(325)$ using Euler's product formula for ϕ.

Problem 7.12 Calculate $\phi(1914)$ using Euler's product formula for ϕ.

Problem 7.13 Find a reduced set of residue classes modulo 20. What is $\phi(20)$?

Problem 7.14 What is the multiplicative inverse of 13 modulo 20?

Problem 7.15 Evaluate $13^7 \pmod{20}$.

Problem 7.16 Find the last two digits of 13^{1323}.

Problem 7.17 Find the last two digits of 35^{2017}.

Problem 7.18 What is the smallest $x > 0$ such that $x \equiv 7 \pmod 8$ and $x \equiv 11 \pmod{12}$?

Problem 7.19 Does the Chinese Remainder Theorem apply to finding a solution to $x \equiv 7 \pmod 8$ and $x \equiv 11 \pmod{12}$?

Problem 7.20 Find the smallest $x > 0$ such that

$$
\begin{aligned}
x &\equiv 1 \quad (\text{mod } 3) \\
x &\equiv 3 \quad (\text{mod } 5) \\
x &\equiv 5 \quad (\text{mod } 7).
\end{aligned}
$$

7.3 Practice Questions

Problem 7.21 Calculate $\phi(60)$ and find a reduced residue set modulo 60.

Problem 7.22 Calculate the following using Euler's product formula for ϕ.

(a) Suppose p is a prime. What is $\phi(p^k)$?

(b) Suppose p and q are distinct primes. What is $\phi(p^k \cdot q^l)$?

(c) Calculate $\phi(400)$.

Problem 7.23 Find the

(a) last two digits of 3^{2006}.

(b) last three digits of 11^{2002}.

Problem 7.24 Find the remainder of 11^{2015} when it is divided by 19.

Problem 7.25 Use the direct method to find all simultaneous solutions to the system $x \equiv 3 \pmod{5}$ and $x \equiv 4 \pmod{11}$.

Problem 7.26 Proof of the Chinese Remainder Theorem Example

(a) Calculate the modular multiplicative inverse of $11 \cdot 17$ modulo 6.

(b) Calculate the modular multiplicative inverse of $6 \cdot 17$ modulo 11.

(c) Calculate the modular multiplicative inverse of $6 \cdot 11$ modulo 17.

Problem 7.27 Solve the system:

$$x \equiv 5 \pmod 6, \quad x \equiv 4 \pmod{11}, \quad x \equiv 3 \pmod{17}.$$

Problem 7.28 In an earlier problem we wanted a number x such that $x \equiv k \pmod{k+1}$ for $k = 1, 2, 3, 4, 5$.

Show that you can always find an x so that $x \equiv k \pmod{k+1}$ for $k = 1, 2, 3, \ldots, M$ for any integer M.

Problem 7.29 Find the remainder when 9^{9^9} is divided by 7.

Problem 7.30 Suppose $\gcd(m, n) = 1$. Prove that $m^{\phi(n)} + n^{\phi(m)} \equiv 1 \pmod{mn}$.

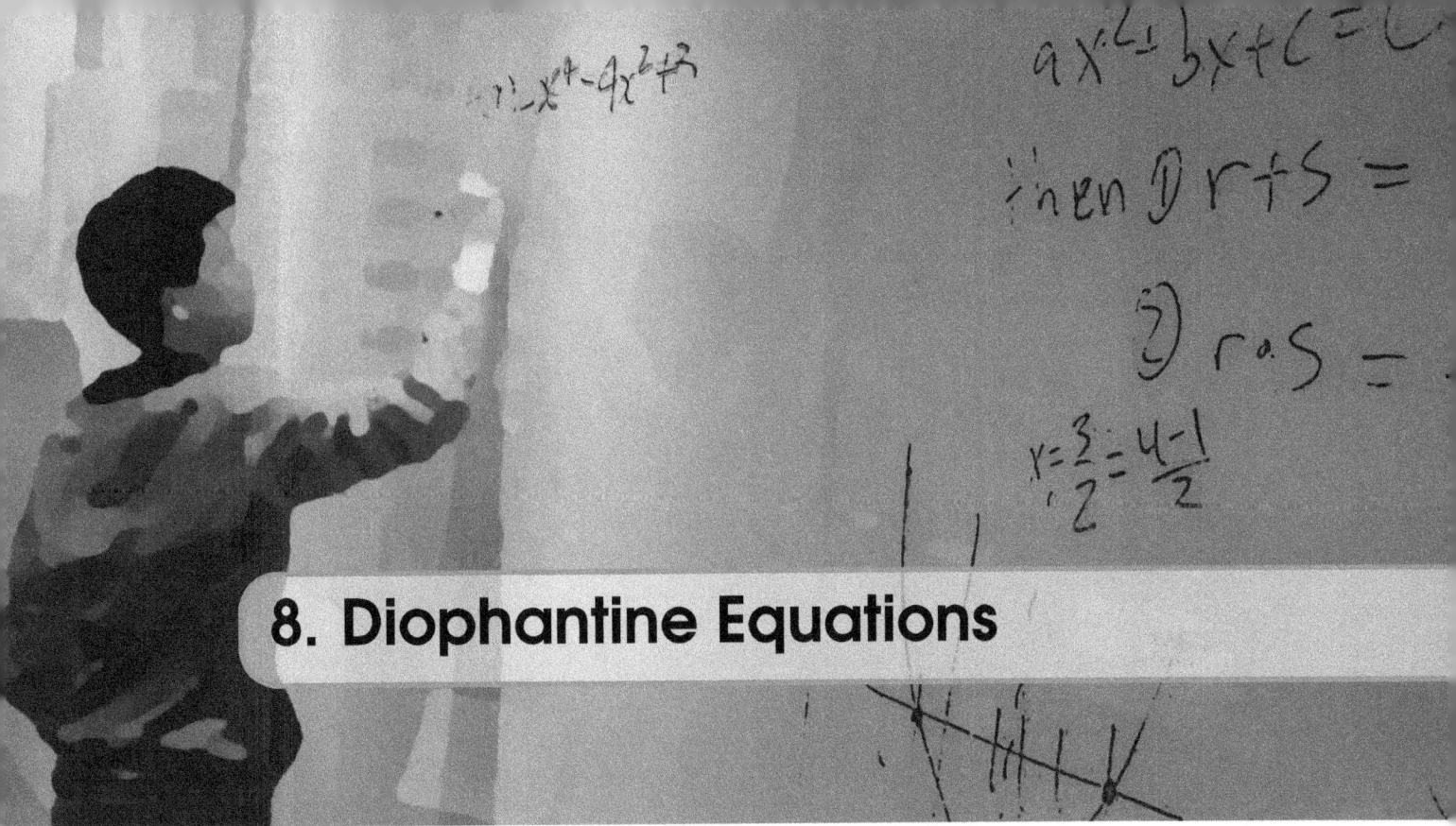

8. Diophantine Equations

Pythagorean Triples

- A *Pythagorean triple* is a triple of positive integers (a, b, c) such that $a^2 + b^2 = c^2$.
- A *primitive Pythagorean triple* is a Pythagorean triple where $\gcd(a, b, c) = 1$.
- For example, $(3, 4, 5)$, $(5, 12, 13)$, $(7, 24, 25)$, $(8, 15, 17)$ are all primitive Pythagorean triples, and $(6, 8, 10)$, $(10, 24, 26)$, $(9, 12, 15)$ are all non-primitive Pythagorean triples.
- The following facts are useful (see the questions for some of the proofs):
  - If a triple of integers (a, b, c) is a Pythagorean triple, then so is (ka, kb, kc) where k is any positive integer.
  - All primitive Pythagorean triples (a, b, c) can be obtained from the formula: $a = m^2 - n^2, b = 2mn, c = m^2 + n^2$ where m, n are integers such that $\gcd(m, n) = 1$.

Diophantine Equations

- A *Diophantine equation* is an equation with integer coordinates, for which we only want integer solutions.
- For example, Pythagorean triples are solutions to the Diophantine equation $a^2 + b^2 = c^2$.
- Diophantine equations are often difficult to solve (and sometimes no efficient method exists). However, in certain cases techniques in number theory can provide ways to find solutions.

8.1 Example Questions

Problem 8.1 Prove the following.

(a) If (a,b,c) is a Pythagorean triple, then (ka,kb,kc) is another Pythagorean triple.

(b) If $m > n$ then $(m^2 - n^2, 2mn, m^2 + n^2)$ is a Pythagorean triple.

(c) If $m - n$ is a positive odd integer with $\gcd(m,n) = 1$, then $(m^2 - n^2, 2mn, m^2 + n^2)$ is a primitive Pythagorean triple.

Problem 8.2 Find all Pythagorean triples containing 25.

Problem 8.3 Suppose (a,b,c) is a primitive Pythagorean triple. Prove that $a \cdot b \cdot c$ is divisible by 60.

Problem 8.4 Solve the following Diophantine equations.

(a) $3x + 4y = 5$.

(b) $3x + 6y = 7$.

(c) $4x + 6y = 14$.

Problem 8.5 Solve the following over the integers.

(a) $x^2 - 7y = 17$.

(b) $x^2 - 7y = 4$.

Problem 8.6 Prove that the equation $\dfrac{1}{x} - \dfrac{1}{y} = \dfrac{1}{p}$ has exactly one solution over the positive integers.

Problem 8.7 Find all solutions to $\dfrac{1}{x} - \dfrac{1}{y} = \dfrac{1}{3}$ if x, y are allowed to be any integers (positive or negative).

Problem 8.8 Solve $3 \cdot 4^m + 1 = n^2$ over the integers.

Problem 8.9 Let n be a positive integer, and $\dfrac{n(n+1)}{2} - 1$ is a prime number. Find all possible values of n.

Problem 8.10 Find all ordered triples (x, y, z) of prime numbers satisfying the equation $x(x+y) = z + 120$.

8.2 Quick Response Questions

Problem 8.11 Are there infinitely many positive integer solutions to $x^2 + y^2 = z^2$? You do not need to prove your answer.

Problem 8.12 Are there infinitely many positive integer solutions to $x^3 + y^3 = z^3$? You do not need to prove your answer.

Problem 8.13 15 is contained in multiple primitive Pythagorean triples. What is the smallest integer that appears with 15 in one of these triples?

Problem 8.14 Consider integer solutions to $2x + 5y = 1$. What is the smallest positive value of x that appears in these solutions?

Problem 8.15 Consider integer solutions to $2x + 5y = 1$. All of the values of x appearing in these solutions are congruent modulo m. What is m?

Problem 8.16 Find all integer solutions to $\dfrac{1}{x} + \dfrac{1}{y} = \dfrac{1}{5}$ with $0 < x < y$. How many pairs of solutions are there?

Problem 8.17 Does the equation $x^2 - 8y^2 = 1$ have integer solutions with $x, y \neq 0$?

Problem 8.18 Does the equation $x^2 - 9y^2 = 1$ have integer solutions with $x, y \neq 0$?

Problem 8.19 Does the equation $x^2 - 8y = 1$ have infinitely many integer solutions?

Problem 8.20 Does the equation $x^2 - 8y = 2$ have infinitely many integer solutions?

8.3 Practice Questions

Problem 8.21 Find all primitive Pythagorean triples $a^2 + b^2 = c^2$ with $c < 50$.

Problem 8.22 Find all Pythagorean triples containing the number 29.

Problem 8.23 What is the largest integer M such that $M \mid a \cdot b \cdot c$ for all Pythagorean triples (a, b, c) (not necessarily primitive)? Prove your answer.

Problem 8.24 Consider the Diophantine equation $Ax + By = C$ with $A, B, C > 0$.

(a) Suppose A, B are fixed. For what values of C does the equation have integer solutions?

(b) For such a C (so the equation does have solutions), describe how to find all the solutions to the equation.

Problem 8.25 Answer the following.

(a) Show that $2x^2 - y^3 = 6$ has no solutions over the integers.

(b) Show that $2x^2 - y^3 = 5$ has solutions over the integers.

Problem 8.26 Find all solutions to $\dfrac{1}{x} - \dfrac{1}{y} = \dfrac{1}{4}$ over the positive integers.

Problem 8.27 Prove that $\dfrac{1}{x} - \dfrac{1}{y} = \dfrac{1}{p}$ always has at least two solutions if we allow x, y to be positive or negative.

Problem 8.28 The following equations each have two solutions for (m, n). Find them.

(a) $3^m + 7 = 2^n$.

(b) $3^m + 5 = 2^n$.

Problem 8.29 Let $f(n) = \dfrac{n(n+1)}{2} - 1$. Prove that $f(n), f(n+1)$ are both even for infinitely many n.

Problem 8.30 Find all ordered triples (x, y, z) of prime numbers satisfying the equation $x(x+y)(y+z) = 140$.

9. Floor Function

The Floor and Fractional Part Functions

- Let x be a real number. Denote $\lfloor x \rfloor$ as the largest integer not exceeding x. We call this function the *floor* or *integer part*.
- For example, $\lfloor 2 \rfloor = 2$, $\lfloor \sqrt{3} \rfloor = 1$, $\lfloor -3.5 \rfloor = -4$.
- For a real number x, define $\{x\} = x - \lfloor x \rfloor$. This function is the *fractional part* of the real number x, therefore $0 \le \{x\} < 1$.
- For example, $\{3\} = 0$, $\{1.25\} = 0.25$, $\{-3.14\} = 0.86$.

Properties of $\lfloor x \rfloor$ and $\{x\}$

- Suppose x, y are real numbers. Then
  - $x = \lfloor x \rfloor + \{x\}$.
  - $x - 1 < \lfloor x \rfloor \le x < \lfloor x \rfloor + 1$.
  - If $x \le y$, then $\lfloor x \rfloor \le \lfloor y \rfloor$.
  - Let n be an integer, then $\lfloor x + n \rfloor = \lfloor x \rfloor + n$.
  - $\lfloor x + y \rfloor \ge \lfloor x \rfloor + \lfloor y \rfloor$.
  - If $\lfloor x \rfloor = \lfloor y \rfloor$ then $|x - y| < 1$.
- Recall that if k is an integer, there are $\left\lfloor \dfrac{n}{k} \right\rfloor$ multiples of k between 1 and n.

9.1 Example Questions

Problem 9.1 Prove the following.

(a) If $x \geq 0$ and $y \geq 0$, then $\lfloor xy \rfloor \geq \lfloor x \rfloor \lfloor y \rfloor$.

(b) Let n be a positive integer. Then $\left\lfloor \dfrac{\lfloor x \rfloor}{n} \right\rfloor = \left\lfloor \dfrac{x}{n} \right\rfloor$.

Problem 9.2 Maximum prime power in $n!$.

(a) Prove that the maximum power of a prime number p in $n!$ is $\left\lfloor \dfrac{n}{p} \right\rfloor + \left\lfloor \dfrac{n}{p^2} \right\rfloor + \left\lfloor \dfrac{n}{p^3} \right\rfloor +$ $\cdots$.

(b) Find the number of zeros at the end of $2016!$.

Problem 9.3 Let $x = \dfrac{1}{3 - \sqrt{7}}$. Find $\lfloor x \rfloor + (1 + \sqrt{7})\{x\}$.

Problem 9.4 Solve the following

(a) $\lfloor x \rfloor^3 - 2\lfloor x^2 \rfloor + \lfloor x \rfloor = 1 - x$.

(b) $\lfloor x^2 \rfloor = \lfloor x \rfloor^2$ for $x \geq 0$.

Problem 9.5 Solve the following equations involving floor functions.

(a) $\lfloor 3x - 1 \rfloor = 2x - \dfrac{1}{2}$. Hint: $\lfloor 3x - 1 \rfloor$ must be an integer, so call it t.

(b) $\left\lfloor \dfrac{5 + 6x}{8} \right\rfloor = \dfrac{15x - 7}{5}$.

Problem 9.6 Find all integers x that satisfy $\lfloor -1.77x \rfloor = \lfloor -1.77 \rfloor x$.

Problem 9.7 Let x be a fraction with $0 < x < 1$. Let $x_1 = x$ and recursively define a_k and x_k by $a_k = \left\lfloor \dfrac{1}{x_k} \right\rfloor + 1$ if $\left\{ \dfrac{1}{x_k} \right\} \neq 0$ and $x_{k+1} = x_k - \dfrac{1}{a_k}$. We'll show in the next problem that

$$x = \frac{1}{a_1} + \frac{1}{a_2} + \cdots + \frac{1}{a_n},$$

where n is the first integer for which $\left\{ \dfrac{1}{x_n} \right\} = 0$, and then $a_n = \dfrac{1}{x_n}$.

Find this expansion, an example of an "Egyptian Fraction" for $x = \dfrac{11}{13}$.

Problem 9.8 Let x and the sequences $\{a_k\}$, and $\{x_k\}$ be as in the previous question.

(a) Suppose $x_k = \dfrac{r}{s}$ and $x_{k+1} = \dfrac{t}{u}$ (where both fractions are reduced). Prove that $0 < t < r$.

(b) Prove that the expansion described always produces a finite Egyptian Fraction.

Problem 9.9 Solve $x + 2\{x\} = 2\lfloor x \rfloor$.

Problem 9.10 $4x^2 - 40\lfloor x \rfloor + 51 = 0$.

9.2 Quick Response Questions

Problem 9.11 Find $\lfloor 9.27 \rfloor + \lfloor -9.27 \rfloor$.

Problem 9.12 Find $\{9.27\} + \{-9.27\}$.

Problem 9.13 What is $\lfloor \sqrt{8} \rfloor$?

Problem 9.14 Evaluate $\lfloor \sqrt{3} \rfloor - \lfloor -\sqrt{3} \rfloor$.

Problem 9.15 Find the maximum power of 3 that is a factor of 30!.

Problem 9.16 Is it true that $\lfloor x+y \rfloor = \lfloor x \rfloor + \lfloor y \rfloor$ for non-integers x, y?

Problem 9.17 Find $\left\lfloor \dfrac{1}{\{2.17\}} \right\rfloor$.

Problem 9.18 Simplify $x = \dfrac{1}{\sqrt{5} - \lfloor \sqrt{5} \rfloor}$ and find $\lfloor x \rfloor$.

Problem 9.19 Find the smallest solution to $\lfloor x \rfloor + 2\{x\} = 5$.

Problem 9.20 Find the smallest solution to $\lfloor x \rfloor \cdot \{x\} = 5$. Input your answer rounded to the nearest hundredth if necessary.

9.3 Practice Questions

Problem 9.21 Prove that $\lfloor 2x \rfloor + \lfloor 2y \rfloor \geq \lfloor x \rfloor + \lfloor y \rfloor + \lfloor x+y \rfloor$ for all real x, y. Hint: Consider the cases that (i) $\{x\}, \{y\} < 1/2$, (ii) $1/2 \leq \{x\}$ and $\{y\} < 1/2$ or $\{x\} < 1/2$ and $1/2 \leq \{y\}$, and (iii) $1/2 \leq \{x\}, \{y\}$.

Problem 9.22 Find the maximum power of 3 in 2016!.

Problem 9.23 Let $x = \dfrac{1}{\sqrt{3} - \sqrt{2}}$. Find $\lfloor x \rfloor + (\sqrt{2} - \sqrt{3})\{x\}$.

Problem 9.24 Find all solutions to $x^2 = \lfloor x \rfloor \cdot \{x\}$.

Problem 9.25 Solve the equation $\left\lfloor \dfrac{5x-2}{5} \right\rfloor = \dfrac{8x-2}{11}$.

Problem 9.26 Suppose x is a real number.

(a) For what values of x is $\lfloor -x \rfloor = -\lfloor x \rfloor$.

(b) What can you say about $\lfloor -x \rfloor$ and $-\lfloor x \rfloor$ is x is not as in part (a)? Prove your answer.

Problem 9.27 Using the method described in the example problems, find an example of an Egyptian Fraction for $\frac{9}{11}$.

Problem 9.28 In general, an Egyptian Fraction is writing a fraction between 0 and 1 in the form

$$\frac{1}{a_1} + \frac{1}{a_2} + \frac{1}{a_3} + \cdots + \frac{1}{a_k}.$$

For example,

$$\frac{7}{12} = \frac{1}{2} + \frac{1}{12} \text{ or } \frac{1}{3} + \frac{1}{4}.$$

The method given in the example problem always produces one such fraction, but it is not always the shortest.

Find an Egyptian Fraction expression for $\frac{5}{91}$ of length 2. That is write $\frac{5}{91} = \frac{1}{m} + \frac{1}{n}$. Show that this is (much) shorter than the fraction produced by the method given above. Hint: $91 = 7 \times 13$ so consider m, n multiples of 7.

Problem 9.29 $\lfloor x \rfloor^2 = x \cdot \{x\}$ where $x > 1$. Hint: Solve for $\{x\}/\lfloor x \rfloor$.

Problem 9.30 Solve $x^2 = \lfloor 3x \rfloor$.

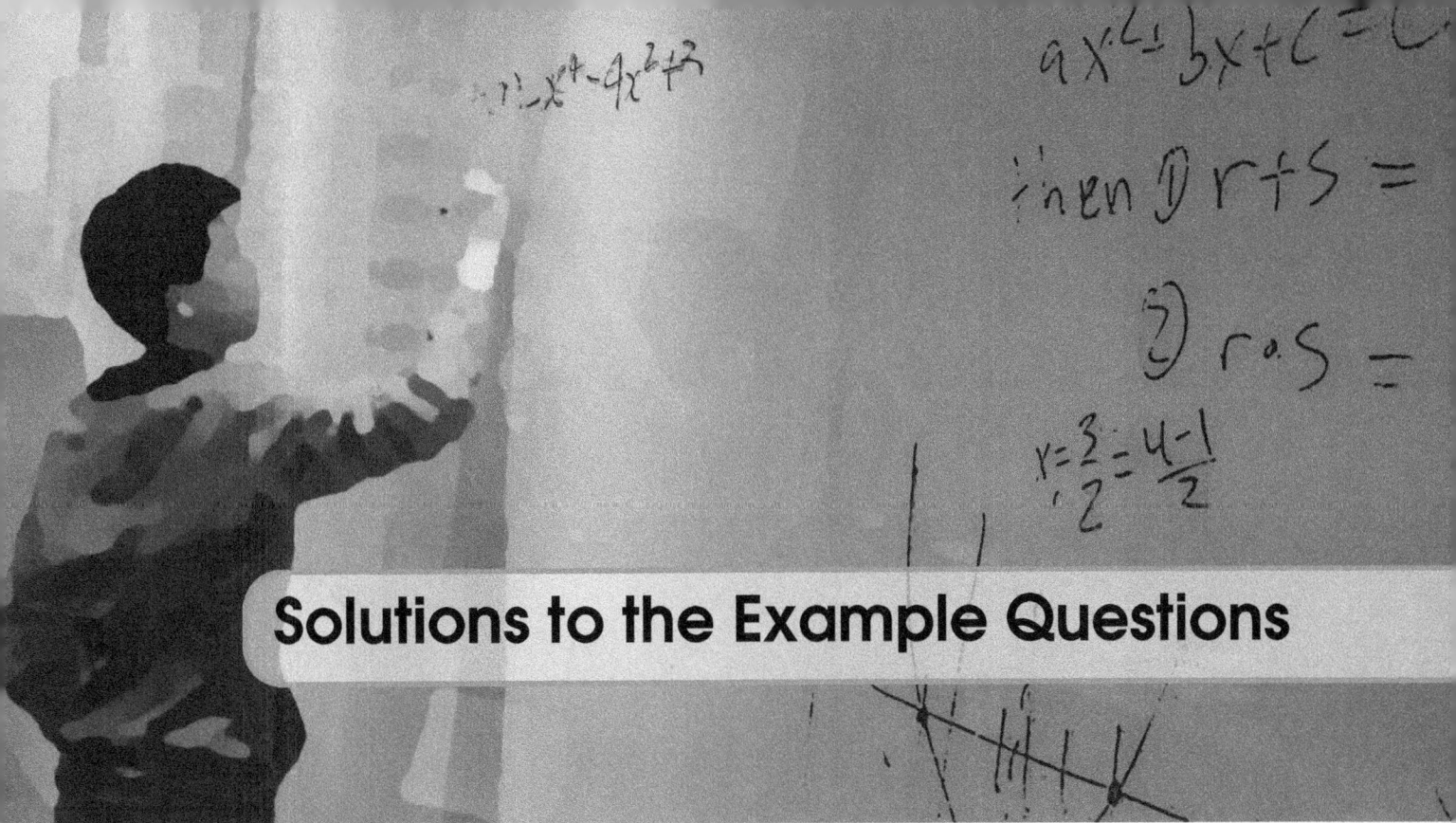

Solutions to the Example Questions

In the sections below you will find solutions to all of the Example Questions contained in this book.

Quick Response and Practice questions are meant to be used for homework, so their answers and solutions are not included. Teachers or math coaches may contact Areteem at info@areteem.org for answer keys and options for purchasing a Teachers' Edition of the course.

1 Solutions to Chapter 1 Examples

Problem 1.1 Convert the following numbers written in different bases to decimal.

(a) 1232_4

Answer

110

Solution

We have
$$1 \cdot 4^3 + 2 \cdot 4^2 + 3 \cdot 4 + 2 = 64 + 32 + 12 + 2 = 110.$$

(b) 10120_3

Answer

96

Solution

We have
$$1 \cdot 3^4 + 1 \cdot 3^2 + 2 \cdot 3 = 81 + 9 + 6 = 96.$$

(c) ABC_{16}

Answer

2748

Solution

We have
$$ABC_{16} = 10 \times 16^2 + 11 \times 16 + 12 = 2748.$$

Problem 1.2 Convert the number 98 in decimal to the following bases:

(a) 7

Answer

200_7

Solution

Note $98 = 2 \cdot 49$ so we have $98 = 200_7$.

(b) 3

Answer

10122_3

Solution

Note
$$98 = 81 + 9 + 3 + 3 + 1 + 1$$
so we have $98 = 10122_3$.

(c) 2

Answer

1100010_2

Solution

Note
$$98 = 64 + 32 + 2$$
so we have $98 = 1100010_2$.

(d) 16

Answer

62_{16}

Solution

Note $98 = 6 \cdot 16 + 2$ so $98 = 62_{16}$.

Problem 1.3 Convert the following. Try to come up with an efficient strategy!

(a) 1234567890 in base 10 to base 100. Note: Think of base 100 as having digits
$\boxed{00}, \boxed{01}, \ldots, \boxed{99}$.

Answer

$\boxed{12}\boxed{34}\boxed{56}\boxed{78}\boxed{90}$

Solution

Note trivially

$$1234567890 = 12 \cdot 100^4 + 34 \cdot 100^3 + 56 \cdot 100^2 + 78 \cdot 100 + 90.$$

(b) 1011010110_2 to base 8.

Answer

1326_8

Solution

One trick to convert from base 2 to base 8 (since $2^3 = 8$) is to split the number into 3-digit segments: $1,011,010,110$. Each segment (thought of as a base 2 number) corresponds to one digit in base 8, so the answer is 1326_8.

(c) 4095 from decimal to hexadecimal

Answer

FFF_{16}

Solution

Note $16^3 = 4096$, so

$$4095 = 4096 - 1 = 1000_{16} - 1_{16} = FFF_{16}.$$

(d) 1333 from octal to ternary.

Answer

$1,000,002_3$

Solution

We have
$$1332_8 = 1 \cdot 512 + 3 \cdot 64 + 3 \cdot 8 + 3 = 731.$$
Since $729 = 3^6$, we have
$$1332_8 = 1,000,000_3 + 2_3 = 1,000,002_3.$$

Problem 1.4 Perform the following operations in the given base. Double check your answer by converting to and from decimal as well.

(a) $122_3 + 201_3$.

Answer

$1100_3 = 36$.

Solution

Doing the operation in base 3, we get 1100_3. Else $122_3 + 201_3 = 17 + 19 = 36 = 1100$.

(b) $1010_2 \times 11_2$.

Answer

$11110_2 = 30$.

Solution

Doing the operation in base 2, we get 11110_2. Else $1010_2 \times 11_2 = 10 \times 3 = 30 = 11110_2$.

Problem 1.5 The following is a procedure for converting from decimal to binary: Starting with a number N, repeatedly divide by two and record the remainder (0 or 1) at each step. After reaching 0, the remainders written in reverse order will give the number in base 2. For example,

$$10 \div 2 = 5 \, R \, 0; \; 5 \div 2 = 2 \, R \, 1; \; 2 \div 2 = 1 \, R \, 0; \; 1 \div 2 = 0 \, R \, 1 \text{ so } 10 = 1010_2.$$

(a) Use the procedure to convert 47 from decimal to binary.

Answer

101111_2

Solution

Repeatedly dividing by 2 gives quotient, remainder pairs $(23, 1), (11, 1), (5, 1), (2, 1), (1, 0), (0, 1)$ so $47 = 101111_2$.

(b) Explain why the procedure works. Hint: You do not need to give a full proof, but at least explain a small example.

Solution

Recall the conversion of 10 to binary. We ended with $1 = 1$. In the previous step we had $2 = 1 \cdot 2 + 0 = 1 \cdot 2 + 0$ after substituting $1 = 1$ (this is useless here, but establishes the pattern). Then we had $5 = 2 \cdot 2 + 1 = (1 \cdot 2 + 0) \cdot 2 + 1 = 1 \cdot 2^2 + 0 \cdot 2 + 1$ (after substituting $2 = 1 \cdot 2 + 0$. Finally, $10 = 5 \cdot 2 + 0 = (1 \cdot 2^2 + 0 \cdot 2 + 1) \cdot 2 + 0 = 1 \cdot 2^3 + 0 \cdot 2^2 + 1 \cdot 2 + 0 = 1010_2$ as needed.

Problem 1.6 Recall $0.375 = 1/4 + 1/8$ so $0.375 = 0.011_2$. Devise a procedure for converting decimals less than 1 to binary. Use your method to convert the following to binary.

(a) 0.3125.

Answer

0.0101_2

Solution

Multiply the decimal by 2, record the ones digit (either 0 or 1) and remove the ones digit. Then repeat. $0.3125 \times 2 = 0.625$ so first decimal digit is 0; $0.625 \times 2 = 1.25$ so second digit is 1; $0.25 \times 2 = 0.5$ so third digit is 0; lastly, $0.5 \times 2 = 1$ so the fourth (and last) digit is 1.

(b) $0.\overline{3}$.

Answer

$0.\overline{01}_2$

Solution

Use a similar procedure to that in part (a). $0.\overline{3} \times 2 = 0.\overline{6}$ (record 0). $0.\overline{6} \times 2 = 1.\overline{3}$ (record 1 and continue with $0.\overline{3}$). Note we returned to $0.\overline{3}$ after recording 01 and that this pattern will continue indefinitely. Hence $1/3 = 0.\overline{3} = 0.\overline{01}_2$.

Problem 1.7 Mystery Bases

(a) Find the value of base b such that the following addition is correct:

$$531_b + 135_b = 1110_b.$$

Answer

6

Solution

Since the digit 5 appears, the base is at least 6. The addition of the rightmost two digits gives 0. This means the base must be 6.

(b) Explain the joke "Halloween equals Christmas".

Solution

31 Oct = 25 Dec, interpreted as "31 in octal (base 8) equals 25 in decimal (base 10)". This is a well-known joke in computer science. The origin can be traced to the Isaac Asimov mystery story "The Family Man", first published in 1976.

Problem 1.8 In the sequence $111, 21, x, 12, y, 10, z, \ldots$ what are x, y, z?

Answer

$x = 13, y = 11, z = 7$

Solution

Note $111_2 = 21_3 = 12_5 = 11_6 = 10_7 = 7$. Hence the sequence is just 7 written in base $2, 3, 4, \ldots$, that is, $111, 21, 13, 12, 11, 10, 7, 7, 7, \ldots$.

Problem 1.9 Based on the digits, how do you tell if a number is even or odd,

(a) if the number is written in base 10?

Answer

Even if and only if last digit is even $(0, 2, 4, 6, 8)$

(b) if the number is written in base 2?

Answer

Even if and only if the last digit is 0

(c) if the number is written in base 3?

Answer

Even if and only if there are an even number of odd digits

Solution

Letting E be even and O be odd, recall $E \cdot E = E \cdot O = E$ and $O \cdot O = O$. As 3 is odd, so are all powers of 3. Any even digits will multiply by a power of 3 to give an even term, so we only need to worry about the odd digits. Then note there must be an even number of these digits, so that the final sum is even.

Problem 1.10 What is the minimum number of weights which enable us to weigh any integer number of grams of gold from 1 to 100 on a standard balance with two pans? You are allowed to use the weights on either side of the pan.

Answer

5.

Solution

The weights needed are $1, 3, 9, 27, 81$. Note that all the numbers between 1 and 100 can be written using 5 digits in base 3: $\overline{abcde}_3 = a \times 81 + b \times 27 + c \times 9 + d \times 3 + e$, where $a, b, c, d, e \in \{0, 1, 2\}$. Since we can use both sides of the balance, we can mimic a digit of 2 using both sides, since $2 = 3 - 1, 2 \times 3 = 9 - 3$, etc. Hence, the 5 weights $1, 3, 9, 27, 81$ suffice. For example, $47 = 81 - 27 - 9 + 3 - 1$.

2 Solutions to Chapter 2 Examples

Problem 2.1 Formally prove the two "easy facts" mentioned earlier. Note: These are not hard to prove, but it is useful to see it formally written out.

(a) If $a \mid n$ and $a \mid m$ then $a \mid n \pm m$.

Solution

By the definition of divisibility, we have $n = k \cdot a, m = j \cdot a$. Hence $n \pm m = (k \pm j) \cdot a$ so $a \mid n \pm m$.

(b) If $a \mid n$ and k is any integer, then $a \mid k \cdot n$.

Solution

By definition, $n = j \cdot a$, so $k \cdot n = (k \cdot j) \cdot a$ so $a \mid k \cdot n$.

Problem 2.2 Prove the divisibility rule for:

(a) 5

Solution

First note that $n = \overline{a_k a_{k-1} \ldots a_j a_{j-1} \ldots a_1 a_0} = \overline{a_k a_{k-1} \ldots a_1} \cdot 10 + a_0$. 5 divides any multiple of 10 so we just need to make sure $5 \mid a_0$ (using our easy facts!). $5 \mid a_0$ if and only if $a_0 = 0$ or 5.

(b) 2^j

Solution

Similar to part (a) note $2^j \mid 10^j$, so if $n = \overline{a_k a_{k-1} \ldots a_j a_{j-1} \ldots a_1 a_0} = \overline{a_k a_{k-1} \ldots a_j} \cdot 10^j + \overline{a_{j-1} \ldots a_1 a_0}$ and the result follows.

(c) 9 when $n = \overline{abcde}$. Note: The proof for the general case is the same idea, but harder to write down neatly.

Solution

We can write $n = \overline{abcde} = 10000a + 1000b + 100c + 10d + e = 9999a + 999b + 99c + $

$9d + a + b + c + d + e$. Thus, $n = (1111a + 111b + 11c + d) \cdot 9 + (a + b + c + d + e)$. Hence, $9 \mid n$ if $9 \mid (a + b + c + d + e)$ as needed.

Problem 2.3 Consider the integer $\overline{2a3a1a}$.

(a) If the number is divisible by 9, what are the possible values for a?

Answer

$a = 1, 4, 7$

Solution

Using the divisibility rule for 9, the integer is divisible by 9 if $9 \mid 6 + 3a$ or $3 \mid 2 + a$. This is only true when $a = 1, 4, 7$.

(b) If the number is divisible by 11, what are the possible values for a?

Answer

$a = 2$

Solution

Using the divisibility rule for 11, the integer is divisible by 11 if $11 \mid 6 - 3a$. This is only true when $6 - 3a = 0$ or $a = 2$.

Problem 2.4 Prove the divisibility rule for 7. That is, prove that $n = 10a + b$ is divisible by 7 if and only if $a - 2b$ is divisible by 7.

Solution

First note

$$7 \mid 10a + b \Leftrightarrow 7 \mid 10a + b - 7a - 7b = 3a - 6b = 3(a - 2b).$$

Now as 3 and 7 are relatively prime, $3(a - 2b)$ is divisible by 7 if and only if $a - 2b$ is divisible by 7 as needed.

Problem 2.5 Possible Numbers

(a) (AIME 1984) The integer n is the smallest positive multiple of 15 such that every digit of n is either 0 or 8. Find n.

Answer

8880

Solution

It has to be a multiple of 5, so the last digit is 0. It is also a multiple of 3, so there have to be three 8s. The smallest such number is 8880.

(b) Find all possible numbers between 5000 and 10000 that are divisible by $2, 3, 5, 9, 11$.

Answer

5940, 6930, 7920, 8910, 9900

Solution

We want numbers that are divisible by $2, 3, 5, 9, 11$, which is equivalent to being divisible by $2 \cdot 5 \cdot 9 \cdot 11 = 990$. Note $990 \cdot 6 = 5940$, is the smallest possible number. The others (all multiples of 990) are $6930, 7920, 8910, 9900$.

Problem 2.6 A five-digit number has five distinct digits, and it is divisible by 9. What is the largest such number?

Answer

98730

Solution

The largest possible number 98765 is not divisible by 9. If we try to change only the last digit, it doesn't work. Hence try the number $\overline{987ab}$. We need $a + b + 7 + 8 + 9 = a + b + 24$ to be divisible by 9, where $a, b \in \{0, 1, 2, 3, 4, 5, 6\}$. From here it is routine to check that $a = 3, b = 0$ yields the largest number that works.

Problem 2.7 Let k be an even number. Is it possible to write 1 as the sum of the reciprocals of k odd integers?

Answer

No.

Solution

We prove this by contradiction: suppose $\dfrac{1}{n_1} + \dfrac{1}{n_2} + \cdots + \dfrac{1}{n_k} = 1$, multiply both sides by the denominators, then

$$n_2 n_3 \cdots n_k + n_1 n_3 \cdots n_k + \cdots n_1 n_2 \cdots n_{k-1} = n_1 n_2 \cdots n_k,$$

however, the left hand side is even, and the right hand side is odd, a contradiction. Hence it is impossible to write 1 as such a sum.

Problem 2.8 Prove that $\gcd(b,a) = \gcd(b-a,a)$. Hint: For integers m,n if $m \mid n$ and $n \mid m$ then $m = n$.

Solution

Let $d = \gcd(b,a)$ and $e = \gcd(b-a,a)$. Then by definition $d \mid b$ and $d \mid a$, so we know $d \mid b-a$. Therefore (one of our earlier facts) $d \mid \gcd(b-a,a) = e$. Similarly, $e \mid b-a, e \mid a$ so $e \mid (b-a)+a = b$. Hence $e \mid \gcd(b,a) = d$. Therefore $d \mid e$ and $e \mid d$ so $d = e$ as needed.

Problem 2.9 Every number less than 100000 can be written as $\overline{abcde}$ where $a,b,c,d,e \in \{0,1,\ldots,9\}$. For how many of these numbers is it true that 11 divides the sum of $\overline{abc}$ and $\overline{de}$.

Answer

9090.

Solution

$\overline{abcde} = \overline{abc} \times 100 + \overline{de} = \overline{abc} \times 99 + \overline{abc} + \overline{de}$, so in fact the requirement is simply that the original five-digit number $\overline{abcde}$ is divisible by 11. The number of five-digit numbers divisible by 11 is $\lfloor 99999/11 \rfloor = 9090$.

Problem 2.10 Given a six-digit number $\overline{abcdef}$, whose digits are 1,2,3,4,5,6, not necessarily in this order. Assume that $6 \mid \overline{abcdef}, 5 \mid \overline{abcde}, 4 \mid \overline{abcd}, 3 \mid \overline{abc}$, and $2 \mid \overline{ab}$. Find $\overline{abcdef}$.

Solution

123654, 321654. It can be determined immediately that $e = 5$. Also we know that b,d,f must be even digits, and so a,c are odd. So either $a = 1, c = 3$ or $c = 1, a = 3$. To make

$\overline{abc}$ a multiple of 3, only $b = 2$ works. To determine d, note that we only need to have $4 \mid \overline{cd}$, and it could be 16 or 36, but not 14 or 34. So $d = 6$ and then $f = 4$. Hence there are two possible solutions.

3   Solutions to Chapter 3 Examples

Problem 3.1 Prove the following:

(a) The "number of factors" formula above.

Solution

Every factor is of the form $p_1^{f_1} p_2^{f_2} \cdots p_k^{f_k}$ with each f_i between 0 and e_i (inclusive). Hence there are $e_i + 1$ choices for each f_i. Therefore the total number of factors is $(e_1 + 1)(e_2 + 1) \cdots (e_k + 1)$ as needed.

(b) A number has an odd number of factors if and only if it is a square.

Solution 1

Note all the factors come in pairs. For any number n, if a is a factor, so is n/a. Pairing up the factors in this way, we see that a number has an even number of factors unless $a = \sqrt{n}$ is a factor (which is then paired up with itself).

Solution 2

If $n = p_1^{e_1} p_2^{e_2} \cdots p_k^{e_k}$, we know n has $(e_1 + 1) \cdots (e_k + 1)$ factors. Note if one of these terms (the $(e_i + 1)$'s) is even, then n has an even number of factors. Else all the terms are odd, so e_i is even for all i. Hence n is a perfect square.

Problem 3.2 Find the smallest positive integer n such that

(a) $\sqrt{200n}$ is an integer.

Answer

2

Solution

Since $200 = 2^3 \cdot 5^2$, we need one more 2 to make the product a square.

(b) $\sqrt{200n}$ is a perfect cube.

Answer

5000.

Solution

For $\sqrt{200n}$ to be a perfect cube, we need $200n$ to be a power of 6. Recall that $200 = 2^3 \cdot 5^2$. Hence we need three more 2's and four more 5's. The smallest n is therefore $2^3 \cdot 5^4 = 5000$.

Problem 3.3 Compute the product of all distinct positive divisors of 120^6 (express your answer as a power of 120).

Answer

120^{2793}

Solution

Recall that all factors of n come in pairs $a, n/a$, each with product n. The only possible non-paired factor is the square root of a perfect square. For 120^6, it is a perfect square. The prime factorization $120^6 = (2^3 \cdot 3 \cdot 5)^6 = 2^{18} \cdot 3^6 \cdot 5^6$, so there are $(18+1)(6+1)(6+1) = 931$ factors, with 465 pairs (each multiplying to 465) and one square root (which is 120^3). The product of all these factors is $(120^6)^{465} \cdot 120^3 = 120^{2793}$.

Problem 3.4 Prove the following:

(a) $m \cdot n = \gcd(m,n) \cdot \text{lcm}(m,n)$.

Solution

Let $m = p_1^{e_1} p_2^{e_2} \cdots p_k^{e_k}$ and $n = p_1^{f_1} p_2^{f_2} \cdots p_k^{f_k}$. Then

$$m \cdot n = p_1^{e_1+f_1} p_2^{e_2+f_2} \cdots p_k^{e_k+f_k} = \prod_{i=1}^{k} p_i^{e_i+f_i}.$$

We also know that

$$\gcd(m,n) \cdot \text{lcm}(m,n) = \prod_{i=1}^{k} p_i^{\min(e_i,f_i)} \times \prod_{i=1}^{k} p_i^{\max(e_i,f_i)} = \prod_{i=1}^{k} p_i^{\min(e_i,f_i)+\max(e_i,f_i)}.$$

Note that $\min(e_i, f_i) + \max(e_i, f_i) = e_i + f_i$ because either $e_i = f_i = \min(e_i, f_i) = \max(e_i, f_i)$

or $e_i \neq f_i$ and then one of e_i, f_i is the min and the other is the max. Hence $m \cdot n = \gcd(m,n) \cdot \text{lcm}(m,n)$ as needed.

(b) If $a \mid m$ and $a \mid n$, then $a \mid \gcd(m,n)$.

Solution

Let $m = p_1^{e_1} p_2^{e_2} \cdots p_k^{e_k}$ and $n = p_1^{f_1} p_2^{f_2} \cdots p_k^{f_k}$. Then since a is a factor of each, we know $a = p_1^{h_1} p_2^{h_2} \cdots p_k^{h_k}$ where $h_i \leq e_i$ and $h_i \leq f_i$. Hence $h_i \leq \min(e_i, f_i)$ and thus a is a factor of $\gcd(m,n) = \prod_{i=1}^{k} p_i^{\min(e_i, f_i)}$ as needed.

Problem 3.5 Do the following:

(a) Find the greatest common divisor of $2^{2016} - 1$ and $2^{100} - 1$.

Answer

15

Solution

Note that $2^{2016} - 1 = 2^{1916} \times (2^{100} - 1) + (2^{1916} - 1)$ so $\gcd(2^{2016} - 1, 2^{100} - 1) = \gcd(2^{1916} - 1, 2^{100} - 1)$. In general, note that (for $m > n$), $2^m - 1 = 2^{m-n} \times (2^n - 1) + 2^{m-n} - 1$, so $\gcd(2^m - 1, 2^n - 1) = \gcd(2^n - 1, 2^{m-n} - 1)$. Hence, continuing in this manner, we get $\gcd(2^{2016} - 1, 2^{100} - 1) = \gcd(2^{2016-200\cdot100} - 1, 2^{100} - 1) = \gcd(2^{16} - 1, 2^{100} - 1)$. We continue this process to get that the greatest common divisor is equal to $\gcd(2^{16} - 1, 2^{100-6\cdot16} - 1) = \gcd(2^{16} - 1, 2^4 - 1) = \gcd(2^4 - 1, 2^4 - 1) = 15$. **Note:** In fact it is true $m, n \geq 1$ that $\gcd(2^m - 1, 2^n - 1) = 2^{\gcd(m,n)} - 1$.

(b) Show that the fraction
$$\frac{15n + 4}{3n + 1}$$
is irreducible for all positive integers n.

Solution

Note a fraction is irreducible if the numerator and denominator are relatively prime. Use the Euclidean algorithm to get the GCD of the numerator and denominator: $\gcd(15n + 4, 3n + 1) = \gcd(3n + 1, 3n) = \gcd(3n, 1) = 1$.

Problem 3.6 Application of Bezout's Identity

(a) Let $d = \gcd(m,n)$. Prove that if $k = am + bn$ for any integers a, b, then $d \mid k$. Hint: This is not hard, don't overthink!

Solution

By definition of greatest common divisor, $d \mid m$ and $d \mid n$. Thus $d \mid am, d \mid bn$ so $d \mid am + bn = k$.

(b) Suppose a society only has bills of value 34 and 62. Suppose everyone in the society always carries around plenty of both bills. Find all (integer value) prices that it is possible to purchase in this society.

Answer

All multiples of 2

Solution

Note $\gcd(34, 62) = 2$ (either use Euclidean Algorithm or the prime factorization of each number). Therefore Bezout's Identity, it is possible to purchase something with price 2 (so then also any multiple of 2). (In fact, $11 \times 34 - 6 \times 62 = 2$. Note we do NOT actually need to calculate this for the problem!) By part (a), there are no other prices possible.

Problem 3.7 Find all numbers n less than 50 with the following property: the product of the divisors of n is equal to n^2.

Answer

1, 6, 8, 10, 14, 15, 21, 22, 26, 27, 33, 34, 35, 38, 39, 46.

Solution

The factors come in pairs, each pair having a product of n itself. So, the factors 1 and n is one pair, and there would be another pair if n is not 1. That means exactly 4 factors. Therefore, n is either 1, or the product of 2 primes (pq), or the cube of a prime (p^3). So the possible numbers are: 1, 6, 8, 10, 14, 15, 21, 22, 26, 27, 33, 34, 35, 38, 39, 46.

Problem 3.8 Let x, y be positive integers, $x < y$, and $x + y = 667$. Given that $\dfrac{\text{lcm}(x,y)}{\gcd(x,y)} = 120$. Find all such pairs (x, y).

Answer

$(115, 552)$, $(232, 435)$

Solution

Let $c = \gcd(x, y)$, and $x = ac, y = bc$, where $\gcd(a, b) = 1$, and $a < b$. Then note $\text{lcm}(x, y) = abc$. From the given, $ab = 120$. Also, $x + y = (a + b)c = 667 = 23 \times 29$. There are four cases:
(1) $a + b = 1, c = 667$; this is not possible because a and b should be positive integers.
(2) $a + b = 667, c = 1$; this does not give integer roots for a and b.
(3) $a + b = 23, c = 29$; combined with $ab = 120$ and $a < b$, we have $a = 8, b = 15$, so $x = 232, y = 435$.
(4) $a + b = 29, c = 23$; combined with $ab = 120$ and $a < b$, we have $a = 5, b = 24$, so $x = 115, y = 552$.

Problem 3.9 From the set $\{1, 2, \ldots, 100\}$, select k numbers. What is the minimum value of k such that it is guaranteed to have two numbers that are not relatively prime? Hint: How many prime numbers are there less than 100?

Answer

27

Solution

There are 25 primes under 100. Divide the set $\{1, 2, \ldots, 100\}$ into 26 categories: the number 1 itself is a category, and each of the other categories contain the multiples of one of the primes. It is noted that these categories overlap, but together they cover all the set $\{1, 2, \ldots, 100\}$. If we choose 27 numbers, it is guaranteed that there must be two of the numbers that belong to the same category, so these two numbers are not relatively prime. So 27 is sufficient. However, if we only select 26 numbers, the worst case scenario is that we have chosen all the prime numbers plus the number 1, and all those are pairwise relatively prime. Therefore 27 is the minimum value.

Problem 3.10 (Putnam 2000) Given integers $n, m, n \geq m \geq 1$. Show that $\dfrac{\gcd(m,n)}{n}\dbinom{n}{m}$

is an integer. Hint: Recall $\dbinom{n}{m} = \dfrac{n!}{m!(n-m)!}$.

Solution

Use Bezout's identity: There exist integers a, b so that $\gcd(m,n) = am + bn$. Thus

$$
\begin{aligned}
\frac{\gcd(m,n)}{n}\binom{n}{m} &= \frac{am+bn}{n}\binom{n}{m} \\
&= \frac{am}{n}\binom{n}{m} + b\binom{n}{m} \\
&= \frac{am}{n}\cdot\frac{n!}{m!(n-m)!} + b\binom{n}{m} \\
&= a\frac{(n-1)!}{(m-1)!(n-m)!} + b\binom{n}{m} \\
&= a\binom{n-1}{m-1} + b\binom{n}{m}
\end{aligned}
$$

Since binomial coefficients are all integers, the original expression is an integer. Note: The above calculation can be made simpler by recalling the "reduction identity" for binomial coefficients: $k\cdot\dbinom{n}{k} = n\cdot\dbinom{n-1}{k-1}$.

4   Solutions to Chapter 4 Examples

Problem 4.1 Warmups

(a) Prove the equivalence mentioned in the beginning of the packet: $m \mid (a - b)$ if and only if a and b have the same remainder when divided by m.

Solution

By the division algorithm, we can write $a = qm + r$ and $b = pm + s$ where q, p are quotients and r, s are remainders. Therefore $a - b = (q - p)m + (r - s)$. Therefore, $m \mid (a - b)$ if and only if $m \mid r - s$. If $r = s$, then $r - s = 0$ and $m \mid r - s$ as needed. Since $0 \leq r, s < m$ we have $-m < r - s < m$, so if $m \mid (r - s)$ it must be the case that $r - s = 0$, or $r = s$. This completes the proof.

(b) Prove that any year (including a leap year) must have at least one "Friday the 13th".

Solution

Since Friday is 5 days after Sunday, and $13 - 5 \equiv 8 \equiv 1 \pmod{7}$, note that if a month has a "Sunday the 1st" then it has a "Friday the 13th". First assume it is not a leap year. Jan. 1st is day 1 of the year, Feb. 1st the 32 day of the year, etc. As a list the 1st of each of the 12 months is respectively the $(1, 32, 60, 91, 121, 152, 182, 213, 244, 274, 305, 335)$ day of the year. Mod 7, these are all equivalent to $(1, 4, 4, 0, 2, 5, 0, 3, 6, 1, 4, 6)$ mod 7. (Alternatively, $4 \equiv 1 + 31 \pmod 7, 4 \equiv 4 + 28 \pmod 7, 0 \equiv 4 + 31 \pmod 7$, etc.) Note that all of $0, 1, 2, 3, 4, 5, 6$ are included in the list, so every possible day of the week will be the 1st of *some* month. Hence there will be a "Friday the 13th" as needed. Note we proceed identically if it is a leap year. In this case or initial list of days of the year for the 1st of every month is $(1, 32, 61, 92, 122, 153, 183, 214, 245, 275, 306, 336)$, which are equivalent to $(1, 4, 5, 1, 3, 6, 1, 4, 0, 2, 5, 0)$ mod 7. Again each of $0, 1, 2, 3, 4, 5, 6$ appears in the list as needed.

Problem 4.2 Assume that $a \equiv b \pmod m$ and $c \equiv d \pmod m$. Are the following true or false? If false, come up with a counterexample. If true, you'll prove it on your homework!

(a) If $b \equiv c \pmod m$ then $a \equiv c \pmod m$.

Solution

True.

(b) $(a+c) \equiv (b+d) \pmod{m}$.

Solution

True.

(c) $(a \cdot c) \equiv (b \cdot d) \pmod{m}$.

Solution

True.

(d) If k is an integer and $k \mid a, k \mid b$, then $(a/k) \equiv (b/k) \pmod{m}$.

Answer

False

Solution

Let $a = 10, b = 14, m = 4$ (so $a \equiv b \pmod 4$). However, if $k = 2$, then $a/k = 5, b/k = 7$ and $5 \not\equiv 7 \pmod 4$.

(e) If n is a positive integer, then $a^n \equiv b^n$.

Solution

True.

Problem 4.3 Calculations

(a) What is the units digit of 2^{2016}?

Answer

6

Solution

Since we want the units digit, we work mod 10. We have $2^1 = 2$. $2^2 = 2 \cdot 2^1 = 2 \cdot 2 = 4$. $2^3 = 2 \cdot 4 = 8$. $2^4 = 2 \cdot 8 = 16 \equiv 6 \pmod{10}$. $2^5 \equiv 2 \cdot 6 \equiv 2 \pmod{10}$. Hence the units digit if 2^n repeats every fourth power. Since $2016 \equiv 0 \equiv 4 \pmod 4$, $2^{2016} \equiv 2^4 \equiv 6 \pmod{10}$.

(b) Find the remainder when $31^{999} + 65^{100}$ is divided by 32.

Answer

0

Solution

Note that $31 \equiv -1 \pmod{32}, 65 \equiv 1 \pmod{32}$, so $31^{999} + 65^{100} \equiv (-1)^{999} + 1^{100} \equiv -1 + 1 \equiv 0 \pmod{32}$, so the remainder is 0.

(c) What is the units digit of $1^2 + 2^2 + 3^2 + \cdots + 99^2$?

Answer

0

Solution

Note that $\overline{a0}^2 + \overline{a1}^2 + \overline{a2}^2 + \cdots + \overline{a9}^2 \equiv 1^2 + 2^2 + \cdots + 9^2 \pmod{10}$ (for example, $20^2 + 21^2 + \cdots 29^2 = 0^2 + 1^2 + \cdots + 9^2$). Hence, $1^2 + 2^2 + \cdots 99^2 \equiv 10 \times (1^2 + 2^2 + \cdots + 9^2) \equiv 0 \pmod{10}$ so the units digit is 0.

Alternatively, we have $1^2 + 2^2 + \cdots + 99^2 = \dfrac{99 \times 100 \times 199}{6} = 328350$, so the units digit is 0.

Problem 4.4 Answer the following.

(a) A certain natural number n has a unit digit 9 when expressed in base 12. Find the remainder when n^2 is divided by 6.

Answer

3

Solution

We have $n = a_n \times 12^n + a_{n-1} \times 12^{n-1} + \cdots + a_1 \times 12 + a_0$, or less specifically, $n = 12k + 9$ for some integer k. Therefore $n \equiv 9 \equiv 3 \pmod 6$, and hence $n^2 \equiv 9 \equiv 3 \pmod 6$, so the remainder is 3 when n^2 is divided by 6.

(b) If $m > 1$ and $69 \equiv 90 \equiv 125 \pmod m$, what is m?

Answer

7

Solution

Note $90 - 69 = 21, 125 - 90 = 35$, so $m \mid \gcd(21, 35) = 7$. Since $m > 1$ (and 7 is prime) m must be 7.

Problem 4.5 Let $n = \overline{a_k a_{k-1} \ldots a_1 a_0} = a_k 10^k + a_{k-1} 10^{k-1} + \cdots + a_1 10 + a_0, a_i \in \{0, 1, \ldots, 9\}$. Prove the following. Note: These are similar to things you've already proven, but practice using modular arithmetic here!

(a) Prove that $n \equiv \overline{a_{j-1} a_{j-2} \ldots a_1 a_0} \pmod{2^j}$.

Solution

Since $2^j \mid 10^j$, we have (using the properties of modular arithmetic proved above) $n = a_k 10^k + a_{k-1} 10^{k-1} + \cdots + a_1 10 + a_0 \equiv a_{j-1} 10^{j-1} + a_{j-2} 10^{j-2} + \cdots + a_0 \equiv \overline{a_{j-1} a_{j-2} \ldots a_1 a_0} \pmod{2^j}$.

(b) Prove that $n \equiv (a_k + a_{k-1} + \cdots + a_1 + a_0) \pmod 9$. Note this means that a number is equal to the sum of its digits modulo 9.

Solution

Since $10 \equiv 1 \pmod 9$, we have (again using the properties of modular arithmetic proved above) $n = a_k 10^k + a_{k-1} 10^{k-1} + \cdots + a_1 10 + a_0 \equiv a_k 1^k + a_{k-1} 1^{k-1} + \cdots + a_0 \equiv a_k + a_{k-1} + \cdots + a_0 \pmod 9$.

Problem 4.6 Suppose you create a 15 digit number using five 1's, five 2's, and five 3's. Is it possible that your number is a perfect square?

Answer

No

Solution

Note the sum of the digits of the number is equal to $5 + 10 + 15 = 30$. Therefore the number is divisible by 3 (since $3 \mid 30$). However, the number is not divisible by 9 (since $9 \nmid 30$), so the number cannot be a square.

Problem 4.7 The number 2^{29} is a nine-digit number all of whose digits are distinct. Without computing the actual number, determine which of the ten digits is missing.

Answer

4

Solution

$2^{29} \equiv 5 \pmod 9$ using patterns. If all digits were present, the sum of digits would have been $0 + 1 + \cdots + 9 = 45$, a multiple of 9. One digit is missing and we get 5 mod 9, so it is 4 that's missing.

Problem 4.8 The Fibonacci sequence is defined by $F_1 = F_2 = 1$, and $F_{n+2} = F_{n+1} + F_n$, that is, the first two terms are both 1, and each subsequence term is the sum of the previous two terms. Find the remainder when F_{2016} is divided by 7.

Answer

0

Solution

Calculate the terms in mod 7, the pattern repeats every 16 terms:

$$1, 1, 2, 3, 5, 1, 6, 0, 6, 6, 5, 4, 2, 6, 1, 0, 1, 1, \ldots.$$

Since $2016 \equiv 0 \equiv 16 \pmod{16}$ we want the 16th term.

Problem 4.9 What is the smallest five-digit integer divisible by both 8 and by 9?

Answer

10008

Solution

If an integer is divisible by both $8, 9$ it is divisible by $72 = \text{lcm}(8, 9)$. We have $\lfloor \frac{10000}{72} \rfloor = 138$, so the smallest number is $139 \times 72 = 10008$.

Alternatively, 10000 is the smallest such integer. To be divisible by 8, the last three digits must be divisible by 8. 000 works, but $10000 \equiv 1 + 0 + 0 + 0 + 0 \equiv 1 \pmod 9$. The next to test is 008, which works as $10008 \equiv 1 + 0 + 0 + 0 + 8 \equiv 9 \equiv 0 \pmod 9$.

Problem 4.10 There are two two-digit numbers whose square ends in the same two-digit number. Find them.

Answer

25 and 76

Solution

The ones digit has to be a 5 or 6. If the number is $10x + 5$, then $(10x + 5)^2 = 100x^2 + 100x + 25$, so the last two digits are 25. If the number is $10x + 6$, then $(10x + 6)^2 = 100x^2 + 120x + 36 = 100(x^2 + x) + 20x + 30 + 6$. Now since $20x + 30 \equiv 10x \pmod{100}$, then $10x + 30 \equiv 0 \pmod{100}$, which means $x = 7$, so 76 is the other two-digit number that ends in 76 when squared.

5   Solutions to Chapter 5 Examples

Problem 5.1 Modular Multiplicative Inverse

(a) Suppose $\gcd(a,m) = 1$. Consider the set of values $\{0, a, 2a, 3a, \ldots, (m-1)a\}$. Prove that all ($m$) values have distinct remainders when dividing by m.

Solution

Suppose $ka \equiv la \pmod{m}$ for $0 \le l \le k < m$. By definition of congruence, $m \mid (k-l)a$. Since a and m are relatively prime, $m \mid (k-l)$. But we know that $0 \le k - l < m$, so the only possibility is $k - l = 0$ so $k = l$ as needed.

(b) Still assume $\gcd(a,m) = 1$. Prove that there is an integer b with $0 \le b < m$ and $ba \equiv 1 \pmod{m}$. Such a b is called the *modular multiplicative inverse* of a modulo m.

Solution

By part (a), the numbers $0, a, 2a, 3a, \ldots, (m-1)a$ all have distinct remainders when dividing by m. Note that there are m total numbers, and only m possible remainders $(0, 1, 2, \ldots, m-1)$. Since all the numbers have distinct remainders, one of them is equal to 1. Hence $ba \equiv 1 \pmod{m}$ for some $b = 0, 1, 2, \ldots, m-1$ as needed.

Problem 5.2 Prove the following.

(a) If $a \equiv b \pmod{m}$ and $k \mid m$, then $a \equiv b \pmod{k}$.

Solution

$a \equiv b \pmod{m}$ implies that $m \mid (a-b)$. Thus, since $k \mid m$, we have $k \mid (a-b)$ so $a \equiv b \pmod{k}$ as needed.

(b) If $a \equiv b \pmod{m}$, $d \mid a$, $d \mid b$, and $\gcd(d,m) = 1$, then $\dfrac{a}{d} \equiv \dfrac{b}{d} \pmod{m}$.

Solution

By assumption $a = d \cdot e, b = d \cdot f$ for integers e, f. Since $\gcd(d,m) = 1$, there is an integer g such that $g \cdot d \equiv \pmod{m}$. We have $a \equiv b \pmod{m} \Leftrightarrow d \cdot e \equiv d \cdot f \pmod{m}$. Therefore, $g \cdot d \cdot e \equiv g \cdot d \cdot f \pmod{m}$ so $1 \cdot e \equiv 1 \cdot f \pmod{m}$. Since $e = a/d, f = b/d$ this completes the proof.

Problem 5.3 Modular Multiplicative Inverse Practice

(a) Find the modular multiplicative inverse of 8 mod 11.

Answer

7

Solution

Note $7 \cdot 8 = 56 \equiv 1 \pmod{11}$.

(b) Find the modular multiplicative inverse of 8 mod 35.

Answer

22

Solution

Note that $-13 \cdot 8 = -104 \equiv 1 \pmod{35}$. Therefore, since $-13 \equiv 22 \pmod{35}$ we have that 22 is the modular multiplicative inverse of 8 mod 35.

(c) Find the modular multiplicative inverse of 9 mod 70.

Answer

39

Solution

Looking at the units digit, note $9 \cdot 9 = 81$ has units digit 1. Checking $9, 19, 29, 39$ we see that $39 \cdot 9 = 351 \equiv 1 \pmod{70}$.

(d) Find all solutions to $7x \equiv 3 \pmod{11}$.

Answer

$2 + 11k$ for k integers

Solution

Note that 8 is the modular multiplicative inverse of 7, so the equation is equivalent to $x \equiv 3 \cdot 8 \equiv 2 \pmod{11}$. Hence $x \equiv 2 \pmod{11}$, so $x = 2 + 11k$ for integers k.

Problem 5.4 Find the possible remainders of n^2 in

(a) $\pmod 3$.

Answer

0, 1

Solution

Since every integer is equivalent to either $0, 1, 2$ modulo 3, we only need to check $n = 0, 1, 2$. $0^2 \equiv 0 \pmod 3, 1^2 \equiv 1 \pmod 3, 2^2 \equiv 4 \equiv 1 \pmod 3$, so $0, 1$ are the only possibilities.

(b) $\pmod 5$.

Answer

0, 1, 4

Solution

Since every integer is equivalent to either $0, 1, 2, 3, 4$ modulo 5, we only need to check $n = 0, 1, 2, 3, 4$. These numbers have remainder $0, 1, 4, 4, 1$ when divided by 5, so $0, 1, 4$ are the only possibilities.

(c) $\pmod 9$.

Answer

0, 1, 4, 7

Solution

Since every integer is equivalent to either $0, 1, 2, \ldots, 8$ modulo 9, we only need to check $n = 0, 1, 2, \ldots, 8$. These numbers have remainder $0, 1, 4, 0, 7, 7, 0, 4, 1$ when divided by 9, so $0, 1, 4, 7$ are the only possibilities.

Problem 5.5 Let $j \geq 2$ be an integer. If n is an integer, find the possible remainders of n^j in (mod 4).

Answer

$0, 1$ if j is even, $0, 1, 3$ if j is odd.

Solution

Since every integer is equivalent to either $0, 1, 2, 3$ modulo 4, we only need to check $n = 0, 1, 2, 3$. It is clear that $0^j = 0, 1^j = 1$ for all j. Note that $2^2 \equiv 4 \equiv 0$ (mod 4), so $2^j \equiv 0$ (mod 4) for all $j \geq 2$. Finally, $3^2 \equiv 9 \equiv 1$ (mod 4), so $3^j \equiv 1$ if j is even and $3^j \equiv 3$ if j is odd.

Problem 5.6 Let $k \neq 2$ be the product of the first several primes (that is $k = 2 \cdot 3, k = 2 \cdot 3 \cdot 5$, etc.). Prove that *none* of $k, k-1, k+1$ are perfect squares.

Solution

First note that trivially any prime not equal to 2 is odd, hence is $1, 3$ or $1, -1$ mod 4. Recall from above that squares are 0 or 1 mod 4. Since $k \equiv \pm 2 \equiv 2$ (mod 4), k is not a square and $k+1 \equiv 3$ (mod 4) is also not a square. Further, $k \equiv 0$ (mod 3), so $k-1 \equiv -1 \equiv 2$ (mod 3), so $k-1$ is also not a square (as squares are $0, 1$ mod 3).

Problem 5.7 Sum of Squares

(a) Is it possible to find two integers n and m such that $n^2 + m^2 = 2015$?

Answer

No

Solution

Note that $2015 \equiv 3$ (mod 4). We also have from above that n^2, m^2 are either $0, 1$ (mod 4). Hence $n^2 + m^2$ is either $0, 1, 2$ (mod 4). Thus, 2015 is not the sum of two squares.

(b) Prove that there are infinitely many numbers that are not the sum of two squares.

Solution

Consider a number of the form $4k+3$. Then the number is 3 (mod 4). However, as in part (a), the sum of two squares is either $0,1,2$ (mod 4). Hence, any number of the form $4k+3$ is not the sum of two squares.

Problem 5.8 Can a 5-digit number consisting only of distinct even digits be a perfect square?

Answer

No

Solution

Recalling that a number (mod 9) is equal to the sum of its digits, a number consisting of the digits $0,2,4,6,8$ will be equivalent to $0+2+4+6+8 \equiv 2$ (mod 9). However, the only possibilities for squares mod 9 are $0,1,4,7$ from above. Hence any rearrangement of the digits $0,2,4,6,8$ is not a square.

Problem 5.9 Do the following equations have integer solutions?

(a) $x^2 - 5y = 102$.

Answer

No

Solution

Note we have $x^2 = 5y + 102$ so $x^2 \equiv 2$ (mod 5). However, from above this is impossible, so there are no solutions.

(b) $x^2 - 5y = 104$.

Answer

Yes

Solution

Note we have $x^2 = 5y + 104$. Note that $12^2 = 144$, so $x = 12, y = 8$ is a solution. Note, it is not easy to find these solutions, or even clear that they must exist in the general case.

Problem 5.10 Consider the sum of the digits of a perfect square. Is it possible for the sum of the digits to be 2015?

Answer

No

Solution

Note that the sum of the digits of a number is equivalent to the number mod 9. We have $2015 \equiv 8 \pmod 9$, but any perfect square is either $0, 1, 4, 7$ mod 9. Thus, such a number is impossible.

6 Solutions to Chapter 6 Examples

Problem 6.1 Order Sanity Check

(a) Suppose a is an integer and we are working mod m. Prove that the sequence $a, a^2, a^3, a^4, \ldots$ eventually starts repeating.

Solution

We can think of the sequence as a recursive sequence with definition $r_1 = a$ and $r_{n+1} = a \cdot r_n \pmod{m}$. Since we are working mod m, the sequence takes on $\leq m$ different values, so eventually $r_j \equiv r_k \pmod{m}$ for some j, k. Hence $r_{j+1} = r_{k+1}$, etc. so the sequence keeps repeating.

(b) Suppose $\gcd(a, m) = 1$. Show that there is some K (and hence a smallest K) such that $a^K = 1$.

Solution

By part (a), the sequence starts repeating, so there are integers $k < j$ such that $a^k \equiv a^j \pmod{m}$. Since $\gcd(a, m) = 1$, there is a modular multiplicative inverse b of a such that $b \cdot a \equiv 1 \pmod{m}$. Therefore

$$a^k \equiv a^j \pmod{m} \Leftrightarrow b^k \cdot a^k \equiv b^k \cdot a^j \pmod{m} \Leftrightarrow 1 \equiv a^{j-k} \pmod{m}.$$

Therefore setting $K = j - k$ we have $a^K \equiv 1 \pmod{m}$ as needed.

Problem 6.2 Suppose $\gcd(a, m) = 1$. Consider the set $\mathscr{S} = \{a, a^2, a^3, a^4, \ldots\}$.

(a) Fill in the blank to get a true statement: The set S is a complete set of nonzero residue classes modulo m if and only if the multiplicative order of a modulo m is __.

Answer

$m - 1$.

Solution

Note that a complete set of nonzero residue classes modulo m always has size $m - 1$.

(b) Give an example of a prime m so that the set $\{2, 2^2, \ldots\}$ is a complete set of nonzero residue classes modulo m.

Answer

3.

Solution

Answers may vary. Small examples are $m = 3, 5, 11, \ldots$.

(c) Give an example of a prime m so that the set $\{2, 2^2, \ldots\}$ is *not* a complete set of nonzero residue classes modulo m.

Answer

7.

Solution

Answers may vary. Some simple examples are $m = 7, 31, 127 = 2^3 - 1, 2^5 - 1, 2^7 - 1$.

Problem 6.3 Prove Fermat's Little Theorem (if a prime p does not divide a, then $a^{p-1} \equiv 1 \pmod{p}$). Hint: Recall a previous result involving the set $\{a, 2a, 3a, \ldots, (p-1)a\}$.

Solution

We showed that $a, 2a, 3a, \ldots, (p-1)a$ all had distinct remainders after dividing by p, so

$$(p-1)! \cdot a^{p-1} \equiv a \cdot 2a \cdot 3a \cdots \cdots (p-1)a \equiv (p-1)! \pmod{p}.$$

Since $(p-1)!$ has only prime factors that are less than p, $\gcd((p-1)!, p) = 1$, so $(p-1)!$ has a modular multiplicative inverse and we can cancel $(p-1)!$ from both sides to get $a^{p-1} \equiv 1 \pmod{p}$ as needed.

Problem 6.4 Wilson's Theorem

(a) Suppose p is a prime. Show that $x^2 \equiv 1 \pmod{p}$ has exactly 2 solutions modulo p. Hint: Consider the minimal nonnegative complete set of residue classes modulo p.

Solution

$x^2 \equiv 1 \pmod{p}$ is equivalent to $(x-1)(x+1) \equiv 0 \pmod{p}$. Therefore either $p \mid x - 1$ or $p \mid x + 1$. We may work with only the minimal nonnegative complete set of residue classes modulo p, so we can assume $0 \leq x < p$. Hence either $x - 1 = 0$ or $x + 1 = p$ so $x = 1$ or $x = p - 1$ are the only solutions.

(b) Prove Wilson's Theorem (for a prime p, $(p-1)! \equiv -1 \pmod{p}$). Hint: Recall multiplicative inverses.

Solution

The case $p = 2$ is trivial. For $p > 2$, we use a "pairing" method. For each a between 1 and $p - 1$ inclusive, there is a modular multiplicative inverse b. Clearly the inverse of b is also a. Thus a and b is a pair such that $ab \equiv 1 \pmod{p}$. All numbers are paired up except for those values x where $x^2 \equiv 1 \pmod{p}$ (by part (a) the only such x are 1 and $p - 1$). So among $1, 2, 3, \ldots, p - 1$, all are paired up except for 1 and $p - 1$, so $(p-1)! \equiv p - 1 \equiv -1 \pmod{p}$.

Problem 6.5 Compute the following.

(a) The remainder when 9^{2016} is divided by 11.

Answer

9.

Solution

Since 11 is prime, we can use Fermat's Little Theorem to get $9^{10} \equiv 1 \pmod{11}$. Therefore, $9^{2016} \equiv 9^6 \equiv (-2)^6 \equiv 64 \equiv 9 \pmod{11}$.

(b) The remainder of $2^{50} + 3^{50}$ when divided by 13.

Answer

0.

Solution

Apply the Fermat's Little Theorem, $2^{12} \equiv 1 \pmod{13}$, so $2^{50} \equiv 2^2 \equiv 4 \pmod{13}$. Similarly, $3^{50} \equiv 3^2 \equiv 9 \pmod{13}$ (as $50 \equiv 2 \pmod{12}$). Therefore, the remainder is $4 + 9 \equiv 0 \pmod{13}$.

Problem 6.6 One Application of Wilson's Theorem

(a) Find the modular multiplicative inverse of 2 mod 67.

Answer

34.

Solution

Clearly $2 \cdot 34 \equiv 68 \equiv 1 \pmod{67}$, so 34 is the modular multiplicative inverse mod 67.

(b) Calculate 64! $\pmod{67}$.

Answer

33.

Solution

By Wilson's Theorem we have $66! \equiv -1 \pmod{67}$. Note that $66! \equiv 66 \cdot 65 \cdot 64! \equiv 2 \cdot 64! \pmod{67}$. Therefore, $2 \cdot 64! \equiv -1 \pmod{67}$, so using part (a) we have $64! \equiv 34 \cdot (-1) \equiv 33 \pmod{67}$.

Problem 6.7 Suppose a sequence has recursive definition $a_1 = k$ (k a positive integer) and $a_{n+1} = 5^{a_n}$. Find the remainder when a_{100} is divided by 11.

Answer

1.

Solution

By Fermat's Little Theorem, $5^{10} \equiv 1 \pmod{11}$. Note that $5^n \equiv 5 \pmod{10}$ for all n, hence $a_n \equiv 5 \pmod{10}$ for all $n \geq 2$. Therefore, $a_{99} = 5 + 10j$ for some integer j. Hence
$$a_{100} \equiv 5^{a_{99}} \equiv 5^{5+10j} \equiv 5^5 \pmod{11} \equiv 1 \pmod{11}.$$
Hence the remainder is 1.

Problem 6.8 Find all primes p such that $p \mid 3^p + 1$.

Answer

$p = 2$.

Solution

By Fermat's Little Theorem, $3^p \equiv 3 \pmod{p}$, so $3^p + 1 \equiv 4 \pmod{p}$. Therefore, the only possibility is $p = 2$.

Problem 6.9 Assume $p = 4k + 1$ is a prime. Prove that $\left[\left(\dfrac{p-1}{2} \right)! \right]^2 \equiv -1 \pmod{p}$.

Hint: Use Wilson's Theorem and try pairing the number r with $p - r$ in $(p-1)!$.

Solution

Using Wilson's Theorem we know $-1 \equiv (p-1)! \pmod{p}$. Following the hint, note that $k \cdot (p - k) \equiv -k^2 \pmod{p}$. Therefore,

$$(p-1)! \equiv (1 \cdot (p-1)) \cdot (2 \cdot (p-2)) \cdots (2k \cdot (2k+1)) \equiv 1^2 \cdot 2^2 \cdots (2k)^2 \equiv \left[\left(\frac{p-1}{2} \right)! \right]^2 \pmod{p}$$

as needed.

Problem 6.10 Show that there are infinitely many n such that $3 \mid n2^n + 1$.

Solution

By Fermat's Little Theorem (or patterns) $2^2 \equiv 1 \pmod{3}$, so $2^j \equiv 1 \pmod{3}$ if j is even. Hence if n is even and $n \equiv (-1) \pmod{3}$ we have $n2^n + 1 \equiv (-1) \cdot 1 + 1 = 0 \pmod{3}$. Note this means that $n = 2, 8, 14, 20, \ldots$ (so $n = 2 + 6k$ for integers k) are all divisible by 3.

7   Solutions to Chapter 7 Examples

Problem 7.1 Reduced Residue System and Euler Totient Function Practice

(a) Calculate $\phi(8)$ using the formula above and double check your answer by giving an example of a reduced set of residue classes modulo 8.

Answer

4

Solution

We have $\phi(8) = 8 \cdot \dfrac{1}{2} \cdot = 4$. A reduced set of residue classes modulo 8 is $\{1,3,5,7\}$.

(b) Calculate $\phi(15)$ using the formula above and double check your answer by giving an example of a reduced set of residue classes modulo 15.

Answer

8

Solution

We have $\phi(15) = 15 \cdot \dfrac{2}{3} \cdot \dfrac{4}{5} = 8$. A reduced set of residue classes modulo 15 is $\{1,2,4,7,8,11,13,14\}$.

(c) Calculate $\phi(2001)$.

Answer

1232

Solution

$$\phi(2001) = 2001 \cdot \frac{2}{3} \cdot \frac{22}{23} \cdot \frac{28}{29} = 1232.$$

Problem 7.2 If $\gcd(m,n) = 1$, then $\phi(m \cdot n) = \phi(m) \cdot \phi(n)$ (such a function is called *multiplicative*). The proof of this result is beyond the scope of this course, but we'll examine the fact a little below.

(a) Show (by calculating each separately) that $\phi(30) = \phi(5) \cdot \phi(6)$. (You may use the formula for ϕ.)

Solution

$\phi(30) = 30 \cdot \dfrac{1}{2} \cdot \dfrac{2}{3} \cdot \dfrac{4}{5} = 8$ and $\phi(5) = 5 - 1 = 4, \phi(6) = 6 \cdot \dfrac{1}{2} \cdot \dfrac{2}{3} = 2$ with $2 \cdot 4 = 8$ as needed.

(b) Show (by calculating each separately) that $\phi(200) = \phi(8) \cdot \phi(25)$. (You may use the formula for ϕ.)

Solution

$\phi(200) = 200 \cdot \dfrac{1}{2} \cdot \dfrac{4}{5} = 80$ and $\phi(8) = 8 \cdot \dfrac{1}{2} = 4, \phi(25) = 25 \cdot \dfrac{4}{5} = 20$ with $4 \cdot 20 = 80$ as needed.

(c) Show that $\phi(200) \neq \phi(20) \cdot \phi(10)$. (You may use the formula for ϕ.)

Solution

$\phi(200) = 200 \cdot \dfrac{1}{2} \cdot \dfrac{4}{5} = 80$ and $\phi(10) = 10 \cdot \dfrac{1}{2} \cdot \dfrac{4}{5} = 4$, while $\phi(20) = 20 \cdot \dfrac{1}{2} \cdot \dfrac{4}{5} = 8$. Note $80 \neq 4 \cdot 8$ so it is important that the numbers are relatively prime.

(d) Prove that $\phi(p \cdot q) = \phi(p) \cdot \phi(q)$ for primes p, q. Do this directly from the definition of ϕ, not using the formula. Hint: There are pq total numbers $\{1, 2, \ldots, pq\}$. How many of them are not in a reduced residue system?

Solution

As in the hint, there are pq numbers $\{1, 2, \ldots, pq\}$. Of them, q are a multiple of the number p, and p are a multiple of the number q. Since the number pq is a multiple of both p and q, we are left with $pq - q - p + 1$ members in the reduced residue system modulo pq. Factoring we have $pq - q - p + 1 = (p-1)(q-1)$, so $\phi(pq) = \phi(p)\phi(q)$ as needed.

Problem 7.3 Find the last two digits of:

(a) 13^{1998}. Hint: What is the modular multiplicative inverse of 13 modulo 100?

Answer

29

Solution

$\phi(100) = 100 \cdot \dfrac{1}{2} \cdot \dfrac{4}{5} = 40$, so $13^{40} \equiv 1 \pmod{100}$. Therefore (since 2000 is a multiple of 40), $13^{2000} \equiv 1 \pmod{100}$. Note that $77 \cdot 13 \equiv 1001 \equiv 1 \pmod{100}$, so 77 is the modular multiplicative inverse of 13 modulo 100. Therefore, $13^{1998} \equiv 77^2 \cdot 13^{2000} \equiv 77^2 \equiv 5929 \equiv 29 \pmod{100}$, so the last two digits are 29.

(b) 32^{1998}.

Answer

24

Solution

Note that $\gcd(32, 100) = 4$, so the extension of Fermat's Little Theorem does not apply. Instead, we calculate $32^2 \equiv 24 \pmod{100}, 32^3 \equiv 68 \pmod{100}, 32^4 \equiv 76 \pmod{100}, 32^5 \equiv 32 \pmod{100}$. Therefore, the pattern repeats every four terms. As $1998 \equiv 2 \pmod 4$, we want the 2nd term, 24.

Problem 7.4 Find the remainder of $13^{3^{13}}$ when it is divided by 17.

Answer

4

Solution

By Fermat's Little Theorem, $13^{16} \equiv 1 \pmod{17}$. Note $\phi(16) = 16 \cdot \frac{1}{2} = 8$, so (using the extension of Fermat's Little Theorem), $3^8 \equiv 1 \pmod{16}$. Therefore, $3^{13} \equiv 3^5 \equiv 3 \pmod{16}$. Thus,

$$13^{3^{13}} \equiv 13^3 \equiv 4 \pmod{17},$$

completing the problem.

Problem 7.5 Solve the system

$$\begin{aligned} x &\equiv 1 \quad (\text{mod } 7) \\ x &\equiv 5 \quad (\text{mod } 16) \end{aligned}$$

Answer

$x = 85 + 112k$ for all k

Solution

We use what we'll call the "direct method". Looking at a list of numbers $\equiv 5 \ (\text{mod } 16)$ we get

$$5, 21, 37, 53, 69, 85, 101, \ldots.$$

Working mod 7 these are equivalent to

$$5, 0, 2, 4, 6, 1, 3, \ldots.$$

(Note the pattern here of $+2$ each time, as $16 \equiv 2 \ (\text{mod } 7)$.) Hence we see that 85 is our solution. As $7 \times 16 = \text{lcm}(7, 16) = 112$, all solutions are of the form $85 + 112k$ for integers k.

Problem 7.6 Consider a general system of three equations:

$$\begin{aligned} x &\equiv b_1 \quad (\text{mod } m_1) \\ x &\equiv b_2 \quad (\text{mod } m_2) \\ x &\equiv b_3 \quad (\text{mod } m_3) \end{aligned}$$

with m_1, m_2, m_3 pairwise relatively prime.

(a) Prove that $m_2 \cdot m_3$ has a modular multiplicative inverse modulo m_1. (Similarly, $m_1 \cdot m_3$ has a modular multiplicative inverse modulo m_2 and $m_1 \cdot m_2$ has a modular multiplicative inverse modulo m_3.)

Solution

We are given that

$$\gcd(m_1, m_2) = \gcd(m_1, m_3) = \gcd(m_2, m_3) = 1.$$

From this it follows that

$$\gcd(m_1, m_2 \cdot m_3) = \gcd(m_2, m_1 \cdot m_3) = \gcd(m_3, m_2 \cdot m_3) = 1.$$

Hence $m_2 \cdot m_3$ has a modular multiplicative inverse modulo m_1, etc.

(b) Let c_1 be the modular multiplicative inverse of m_2m_3 modulo m_1 (this exists by part (a)), and similarly let c_2 be the modular multiplicative inverse of m_1m_3 modulo m_2 and let c_3 be the modular multiplicative inverse of m_1m_2 modulo m_3. Show that

$$x = b_1 \cdot c_1 \cdot m_2m_3 + b_2 \cdot c_2 \cdot m_1m_3 + b_3 \cdot c_3 \cdot m_1m_2$$

is a solution to the system of equations.

Solution

First consider $x \pmod{m_1}$. We have

$$\begin{aligned} x &\equiv b_1 \cdot c_1 \cdot m_2m_3 + b_2 \cdot c_2 \cdot m_1m_3 + b_3 \cdot c_3 \cdot m_1m_2 \\ &\equiv b_1 \cdot c_1 \cdot m_2m_3 + 0 + 0 \\ &\equiv b_1 \cdot (c_1 \cdot m_2m_3) \\ &\equiv b_1 \cdot 1 \equiv b_1 \pmod{m_1}, \end{aligned}$$

as c_1 was chosen to be the modular multiplicative inverse of $m_2m_3 \bmod m_1$.

Identical arguments give $x \equiv b_2 \pmod{m_2}$ and $x \equiv b_3 \pmod{m_3}$ as needed.

Problem 7.7 Solve the system

$$\begin{aligned} x &\equiv 2 \pmod 5 \\ x &\equiv 4 \pmod 9 \\ x &\equiv 6 \pmod{11}. \end{aligned}$$

Answer

$292 + 495k$ for all integers k

Solution 1

We have that $9 \cdot 11 \equiv 99 \equiv 4 \pmod 5$, $5 \cdot 11 \equiv 55 \equiv 1 \pmod 9$, and $5 \cdot 9 \equiv 45 \equiv 1 \pmod{11}$.

Since $4 \cdot 4 \equiv 4 \cdot 99 \equiv 1 \pmod 5$ we have that 4 is the modular multiplicative inverse of 99 modulo 5. 1 is the modular multiplicative inverse of 55 modulo 9 and similarly 1 is the modular multiplicative inverse of 45 modular 11.

Hence, using the formula from 7.6,

$$2 \cdot 4 \cdot 99 + 4 \cdot 1 \cdot 55 + 6 \cdot 1 \cdot 45 \equiv 292 \pmod{495}$$

is a solution to the system. Therefore, the solutions are $292 + 495k$ for all integers k.

Solution 2

Working directly,

$$6, 17, 28, 39, 50, 61, 72, 83, 94, \ldots \equiv 6 \pmod{11}.$$

Of these, 94 is the smallest that is 4 (mod 9). Therefore numbers of the form $94 + 99k$ will be satisfy the restrictions both mod 9 and mod 11. Listing these we have

$$94, 193, 292, \ldots,$$

and we see that $292 \equiv 2 \pmod 5$. Hence all solutions are of the form $292 + 495k$ for integers k.

Problem 7.8 Consider the question: "Find the smallest number that leaves a remainder of 1 when divided by 2, 2 when divided by 3, 3 when divided by 4, 4 when divided by 5, and 5 when divided by 6."

(a) Does the Chinese Remainder Theorem apply to the question?

Answer

No

Solution

4 and 6 are both multiples of 2, so the moduli are not pairwise relatively prime.

(b) Solve the question.

Answer

59

Solution

Every number of the form $5 + 6k$ will have remainder 5 when divided by 6. Checking these, $5, 11, 17, 23, 29$, 29 is the first that also has remainder 4 when dividing by 5. As $\gcd(6, 5) = 30$, every number of the form $29 + 30k$ will satisfy the last two requirements. We then check these numbers, $29, 59$, and see that 59 satisfies all the requirements.

Problem 7.9 Find the remainder when $7^3 + 7^{7^2} + 7^{7^3} + 7^{3^4} + \cdots + 7^{7^{20}}$ is divided by 11.

Answer

5

Solution

First note using Fermat's Little Theorem, $7^{10} \equiv 1 \pmod{11}$. Further, using the extension (or patterns), $7^4 \equiv 1 \pmod{10}$. Therefore,

$$7^7 + 7^{7^2} + 7^{7^3} + 7^{7^4} \equiv 7^{7^5} + \cdots + 7^{7^8}$$
$$\equiv 7^{7^9} + \cdots + 7^{7^{12}}$$
$$\equiv 7^{7^{13}} + \cdots + 7^{7^{16}}$$
$$\equiv 7^{7^{17}} + \cdots + 7^{7^{20}} \pmod{11}$$

Hence $7^7 + 7^{7^2} + 7^{7^3} + \cdots + 7^{7^{20}} \equiv 5 \cdot (7^7 + 7^{7^2} + 7^{7^3} + 7^{7^4}) \pmod{11}$. Using patterns we have $7^7 \equiv 6 \pmod{11}$, $7^{7^2} \equiv 7^9 \equiv 8 \pmod{11}$, $7^{7^3} \equiv 7^3 \equiv 2 \pmod{11}$, and $7^{7^4} \equiv 7^1 \equiv 7 \pmod{11}$. Thus, the final answer is $5 \cdot (6 + 8 + 2 + 7) \equiv 5 \pmod{11}$.

Problem 7.10 Let $p \neq q$ are primes with $\gcd(p-1, q-1) = m$. Set $k = (p-1)(q-1)/m$. Prove that if $\gcd(a, pq) = 1$, then $a^k \equiv 1 \pmod{pq}$.

Solution

By Fermat's Little Theorem

$$a^k = (a^{(p-1)/m})^{q-1} \equiv 1 \pmod{q}, \quad a^k = (a^{(q-1)/m})^{p-1} \equiv 1 \pmod{p}.$$

Since $\gcd(p, q) = 1$, this implies that $a^k \equiv 1 \pmod{pq}$.

8 Solutions to Chapter 8 Examples

Problem 8.1 Prove the following.

(a) If (a,b,c) is a Pythagorean triple, then (ka,kb,kc) is another Pythagorean triple.

Solution

If $a^2 + b^2 = c^2$ then $(ka)^2 + (kb)^2 = k^2(a^2+b^2) = kc^2$ as needed.

(b) If $m > n$ then $(m^2 - n^2, 2mn, m^2 + n^2)$ is a Pythagorean triple.

Solution

We have

$$(m^2 - n^2)^2 + (2mn)^2 = m^4 - 2m^2n^2 + n^4 + 4m^2n^2 = m^4 + 2m^2n^2 + n^4 = (m^2 + n^2)^2$$

as needed.

(c) If $m - n$ is a positive odd integer with $\gcd(m,n) = 1$, then $(m^2 - n^2, 2mn, m^2 + n^2)$ is a primitive Pythagorean triple.

Solution

From part (b) we know it is a Pythagorean triple. $m - n$ is odd, so one of m, n is even and the other is odd. Thus, $m^2 - n^2, m^2 + n^2$ are both odd. Suppose $d \mid (m^2 - n^2)$ and $d \mid (m^2 + n^2)$ (so d is odd). Then $d \mid 2m^2$ and $d \mid 2n^2$ (the sums and differences). As d is odd, $d \mid m^2, d \mid n^2$. However, since $\gcd(m,n) = 1$, $\gcd(m^2,n^2) = 1$ and thus $m^2 - n^2, m^2 + n^2$ are relatively prime. Therefore, $\gcd(m^2 - n^2, 2mn, m^2 + n^2) = 1$ as needed.

Problem 8.2 Find all Pythagorean triples containing 25.

Answer

$(7,24,25), (15,20,25), (25,60,65), (25,312,313)$.

Solution

First assume that the triple is primitive. Thus we have $m > n$ such that either (i) $m^2 - n^2 = 25$, (ii) $2mn = 25$ (clearly impossible), or (iii) $m^2 + n^2 = 25$. In case (i) we have $(m - n)(m + n) = 5^2$. As $m - n = m + n = 5$ is impossible, we must have

$m - n = 1, m + n = 25$ so $m = 13, n = 12$. Hence we get a triple of $(13^2 - 12^2, 2 \cdot 13 \cdot 12, 13^2 + 12^2) = (25, 312, 313)$. In case (iii), we have $m^2 + n^2 = 25$ so $m = 4, n = 3$ and therefore we get the triple $(4^2 - 3^2, 2 \cdot 4 \cdot 3, 4^2 + 3^2) = (7, 24, 25)$.

The other possibility is that the triple is not primitive. As $25 = 5^2$ we therefore look for primitive triples containing 5. A method identical to what was done above gives $(5, 12, 13)$ and $(3, 4, 5)$ as the only primitive triples containing 5. Therefore $(25, 60, 65)$ and $(15, 20, 25)$ are also triples containing 25.

In summary, the only Pythagorean triples containing 25 are $(7, 24, 25)$, $(15, 20, 25)$, $(25, 60, 65)$, and $(25, 312, 313)$.

Problem 8.3 Suppose (a, b, c) is a primitive Pythagorean triple. Prove that $a \cdot b \cdot c$ is divisible by 60.

Solution

We show that one of $a = m^2 - n^2, b = 2mn, c = m^2 + n^2$ is divisible by 3, 4, 5 so the product is divisible by 60.

For 3: if m or $n \equiv 0 \pmod{3}$, then $3 \mid b$ (as $b = 2mn$). Else m and $n \equiv \pm 1 \pmod{3}$, so $m^2 \equiv n^2 \equiv 1 \pmod{3}$ and $a = m^2 - n^2 \equiv 0 \pmod{3}$.

For 4: if m or n is even, then $4 \mid b$ (as $b = 2mn$). Else m and $n \equiv \pm 1 \pmod{4}$, so $m^2 \equiv n^2 \equiv 1 \pmod{4}$ and $a = m^2 - n^2 \equiv 0 \pmod{4}$.

For 5: if m or $n \equiv 0 \pmod{5}$, then $5 \mid b$ (as $b = 2mn$). Else we have m^2 and $n^2 \equiv \pm 1 \pmod{5}$. If $m^2 \equiv n^2 \pmod{5}$, then $a = m^2 - n^2 \equiv 0 \pmod{5}$. Otherwise, $c = m^2 + n^2 \equiv 0 \pmod{5}$.

Problem 8.4 Solve the following Diophantine equations.

(a) $3x + 4y = 5$.

Answer

$x = -1 - 4k, y = 2 + 3k$ for integers k

Solution

First note $x = -1, y = 2$ is a solution. Then as $-4 \cdot 3 + 3 \cdot 4 = 0$, $x = -1 - 4k, y = 2 + 3k$ gives all possible solutions.

(b) $3x + 6y = 7$.

Answer

No solutions exist

Solution

$3x + 6y$ is always $0 \pmod 3$, but $7 \not\equiv 0 \pmod 3$, so no solutions exist.

(c) $4x + 6y = 14$.

Answer

$x = 2 - 3k, y = 1 + 2k$ for integers k

Solution

Note this is equivalent to solving $2x + 3y = 7$. First note $x = 2, y = 1$ is a solution. Then as $-3 \cdot 2 + 2 \cdot 3 = 0$, $x = 2 - 3k, y = 1 + 2k$ gives all possible solutions.

Problem 8.5 Solve the following over the integers.

(a) $x^2 - 7y = 17$.

Answer

No solutions

Solution

Working mod 7, we need $x^2 \equiv 17 \equiv 3 \pmod 7$. However, no square is $3 \pmod 7$.

(b) $x^2 - 7y = 4$.

Answer

$x = 7k + 2, y = 7k^2 + 4k$ or $x = 7k + 5, y = 7k^2 + 10k + 3$ for integers k

Solution

Working mod 7, we see that either $x \equiv 2, 5 \pmod 7$. If $x = 7k + 2$ for some k, then

$(7k+2)^2 - 7y = 4$ so solving for y, $y = 7k^2 + 4k$. Similarly, if $x = 7k+5$ we solve $y = 7k^2 + 10k + 3$.

Problem 8.6 Prove that the equation $\dfrac{1}{x} - \dfrac{1}{y} = \dfrac{1}{p}$ has exactly one solution over the positive integers.

Answer

$x = p-1, y = p(p-1)$

Solution

Clearing denominators and rearranging the equation we have that $p(y-x) = xy$. Therefore, as p is prime, either (i) x is a multiple of p, or (ii) y is a multiple of p.

Note in case (i), $x \geq p$ so $\dfrac{1}{x} \leq \dfrac{1}{p}$ so there are no solutions as $\dfrac{1}{y} > 0$.

In case (ii), assume that $y = kp$ for an integer k. Substituting and solving for x we have

$$x = \frac{kp}{k+1} \Rightarrow k+1 = p \Rightarrow k = p-1$$

as p is a prime and $\gcd(k, k+1) = 1$ for all k. Hence we have that

$$x = \frac{(p-1)p}{p-1+1} = p-1 \text{ and } y = p(p-1)$$

as the unique solution.

Problem 8.7 Find all solutions to $\dfrac{1}{x} - \dfrac{1}{y} = \dfrac{1}{3}$ if x, y are allowed to be any integers (positive or negative).

Answer

(x,y) can be $(-6, -2), (2, 6), (4, -12), (6, -6), (12, -4)$

Solution

Note the equation can be rewritten $3y - 3x = xy$ or after completing the rectangle, $(x-3)(y+3) = -9$. Then $-9 = (1)(-9) = (-1)(9) = (3)(-3) = (-3)(3) = (9)(-1) = (-1)(9)$ leads to 6 solutions to $3y - 3x = xy$. However, one of these is $(0,0)$, which doesn't work as $x, y \neq 0$. The other 5 give the 5 solutions listed above.

Problem 8.8 Solve $3 \cdot 4^m + 1 = n^2$ over the integers.

Answer

$m = 0, n = \pm 2, m = 2, n = \pm 7$

Solution

We must have $n^2 \equiv 1 \pmod{3}$, so n is of the form $n = 3k+1$ or $n = 3k+2$. First assume $n = 3k+1$, so we have $3 \cdot 4^m + 1 = (3k+1)^2$. Simplifying, $4^m = 3k^2 + 2k = k(3k+2)$. Hence, $\pm k$ must itself be a power of 2 and similarly $\pm(3k+2)$ must also be a power of 2. This is only possible if $k = 2, -1$. This gives $n = 7$ (with $m = 2$) and $n = -2$ (with $m = 0$). If $n = 3k = 2$, after simplifying we have $4^m = 3k^2 + 4k + 1 = (3k+1)(k+1)$. As before, both factors must be powers of 2, and we see $k = 0, -3$ are the only solutions. These correspond to solutions $n = 2, m = 0$ and $n = -7, m = 2$.

Problem 8.9 Let n be a positive integer, and $\dfrac{n(n+1)}{2} - 1$ is a prime number. Find all possible values of n.

Answer

2 or 3

Solution

$\dfrac{n(n+1)}{2} - 1 = \dfrac{(n+2)(n-1)}{2}$. If $n = 1$ this expression is 0. If $n = 2$, the expression is 2, a prime; if $n = 3$, the expression is 5, a prime. For $n \geq 4$, the pair $n+2$ and $n-1$ consists of an even number at least 4 and an odd number at least 3, therefore the expression is not a prime. Therefore the only solutions are 2 and 3.

Problem 8.10 Find all ordered triples (x, y, z) of prime numbers satisfying the equation $x(x+y) = z + 120$.

Answer

$(2, 59, 2), (11, 2, 23)$

Solution

Consider the following cases for prime number z: (1) If $z = 2$, $x(x+y) = 122 = 2 \times 61$. Thus $x = 2, y = 59$.

(2) If z is an odd prime, then $x(x+y)$ is an odd number. That means both x and $x+y$ are odd, and so $y = 2$. The equation becomes $x(x+2) = z + 120$, and so $x^2 + x - 120 = z$, so $(x+12)(x-10) = z$. But z is prime, then $x - 10 = 1$, so $x = 11, z = 23$.

9   Solutions to Chapter 9 Examples

Problem 9.1 Prove the following.

(a) If $x \geq 0$ and $y \geq 0$, then $\lfloor xy \rfloor \geq \lfloor x \rfloor \lfloor y \rfloor$.

Solution

Let $m = \lfloor x \rfloor, n = \lfloor y \rfloor, r = \{x\}, s = \{y\}$. Then $xy = (m+r)(n+s) = mn + ms + nr + rs$. Therefore, $\lfloor xy \rfloor = \lfloor mn + ms + nr + rs \rfloor = mn + \lfloor ms + nr + rs \rfloor$. Hence $\lfloor xy \rfloor \geq mn = \lfloor x \rfloor \lfloor y \rfloor$ as needed.

(b) Let n be a positive integer. Then $\left\lfloor \dfrac{\lfloor x \rfloor}{n} \right\rfloor = \left\lfloor \dfrac{x}{n} \right\rfloor$.

Solution

Let $x = nq + r$ where q is an integer and r is a real number satisfying $0 \leq r < n$. Then $\lfloor x \rfloor = nq + \lfloor r \rfloor$ and $\left\lfloor \dfrac{\lfloor x \rfloor}{n} \right\rfloor = \left\lfloor q + \dfrac{\lfloor r \rfloor}{n} \right\rfloor = q$ and $\left\lfloor \dfrac{x}{n} \right\rfloor = \left\lfloor q + \dfrac{r}{n} \right\rfloor = q$ as needed.

Problem 9.2 Maximum prime power in $n!$.

(a) Prove that the maximum power of a prime number p in $n!$ is $\left\lfloor \dfrac{n}{p} \right\rfloor + \left\lfloor \dfrac{n}{p^2} \right\rfloor + \left\lfloor \dfrac{n}{p^3} \right\rfloor + \cdots$.

Solution

From above there are $\left\lfloor \dfrac{n}{p} \right\rfloor$ multiples of p between 1 and n, each contributing (at least) one power of p to $n!$. However, every multiple of p^2 (of which there are $\left\lfloor \dfrac{n}{p^2} \right\rfloor$ between 1 and 1) each contributes (at least) one more power of p to $n!$. We continue in this manner for $p^3, p^4, \ldots$ to get the needed formula.

(b) Find the number of zeros at the end of 2016!.

Answer

502

Solution

Each zero is generated by a pair of factors 2 and 5. Since there are more factors of 2 than factors of 5, the zeros are determined by the maximum power of 5. So the number of zeros is

$$\left\lfloor \frac{2016}{5} \right\rfloor + \left\lfloor \frac{2016}{5^2} \right\rfloor + \left\lfloor \frac{2016}{5^3} \right\rfloor + \left\lfloor \frac{2016}{5^4} \right\rfloor + \left\lfloor \frac{2016}{5^5} \right\rfloor + \cdots$$
$$= 403 + 80 + 16 + 3 + 0 + \cdots$$
$$= 502.$$

Problem 9.3 Let $x = \dfrac{1}{3 - \sqrt{7}}$. Find $\lfloor x \rfloor + (1 + \sqrt{7})\{x\}$.

Answer

5

Solution

$x = \dfrac{1}{3 - \sqrt{7}} = \dfrac{3 + \sqrt{7}}{2}$, and $2 < \sqrt{7} < 3$, so $2 < x < 3$, thus $\lfloor x \rfloor = 2$, $\{x\} = x - 2 = \dfrac{\sqrt{7} - 1}{2}$. Therefore $\lfloor x \rfloor + (1 + \sqrt{7})\{x\} = 2 + (1 + \sqrt{7}) \cdot \dfrac{\sqrt{7} - 1}{2} = 2 + 3 = 5$.

Problem 9.4 Solve the following

(a) $\lfloor x \rfloor^3 - 2\lfloor x^2 \rfloor + \lfloor x \rfloor = 1 - x$.

Answer

1

Solution

x must be an integer, so we can get rid of the floor functions: $x^3 - 2x^2 + x = 1 - x$, which is $x^3 - 2x^2 + 2x - 1 = 0$, so $(x - 1)(x^2 - x + 1) = 0$. The only real solution is $x = 1$.

(b) $\lfloor x^2 \rfloor = \lfloor x \rfloor^2$ for $x \geq 0$.

Answer

x such that $n \leq x < \sqrt{n^2 + 1}$ for non-negative integers n.

Solution

Let $n = \lfloor x \rfloor$, then plugging into the equation we get $\lfloor x^2 \rfloor = n^2$, so $n^2 \leq x^2 < n^2 + 1$, hence $n \leq x < \sqrt{n^2 + 1}$ (here n can be any nonnegative integer).

Problem 9.5 Solve the following equations involving floor functions.

(a) $\lfloor 3x - 1 \rfloor = 2x - \dfrac{1}{2}$. Hint: $\lfloor 3x - 1 \rfloor$ must be an integer, so call it t.

Answer

$-3/4$ and $-5/4$

Solution

Using the substitution $t = 2x - \dfrac{1}{2}$ we get $x = \dfrac{t}{2} + \dfrac{1}{4}$, so $3x + 1 = \dfrac{3t}{2} + \dfrac{7}{4}$. Hence $\left\lfloor \dfrac{3t}{2} + \dfrac{7}{4} \right\rfloor = t$. Based on the properties of the floor function, $t \leq \dfrac{3t}{2} + \dfrac{7}{4} < t + 1$, solve and get $-\dfrac{7}{2} \leq t < -\dfrac{3}{2}$, therefore $t = -2$ or -3. The corresponding values for x are $-\dfrac{3}{4}$ and $-\dfrac{5}{4}$.

(b) $\left\lfloor \dfrac{5 + 6x}{8} \right\rfloor = \dfrac{15x - 7}{5}$.

Answer

$7/15$ and $4/5$

Solution

Similar to part (a), let $t = \left\lfloor \dfrac{5 + 6x}{8} \right\rfloor$. Then $x = \dfrac{5t + 7}{15}$. By definition we have $t \leq \dfrac{5 + 6x}{8} < t + 1$, thus $t \leq \dfrac{10t + 39}{40} < t + 1$. Solving for t, $0 \leq \dfrac{39 - 30t}{40} < 1$, and get $-\dfrac{1}{30} < t \leq \dfrac{13}{10}$. So $t = 0$ or 1. Finally, plug these t values back in to solve for x.

Problem 9.6 Find all integers x that satisfy $\lfloor -1.77x \rfloor = \lfloor -1.77 \rfloor x$.

Answer

$x = 0, 1, 2, 3, 4$

Solution

Since $\lfloor -1.77 \rfloor = -2$, the equation can be simplified to $\lfloor -1.77x \rfloor = -2x$. Also, $\{-1.77x\} = -1.77x - \lfloor -1.77x \rfloor$, which means $\lfloor -1.77x \rfloor = -1.77x - \{-1.77x\}$, so $-1.77x - \{-1.77x\} = -2x$, then $0.23x = \{-1.77x\}$. From $0 \le \{-1.77x\} < 1$, we get $0 \le 0.23x < 1$, so $0 \le x < \dfrac{100}{23}$. Thus the possible values for x are $0, 1, 2, 3, 4$.

Problem 9.7 Let x be a fraction with $0 < x < 1$. Let $x_1 = x$ and recursively define a_k and x_k by $a_k = \left\lfloor \dfrac{1}{x_k} \right\rfloor + 1$ if $\left\{ \dfrac{1}{x_k} \right\} \ne 0$ and $x_{k+1} = x_k - \dfrac{1}{a_k}$. We'll show in the next problem that

$$x = \frac{1}{a_1} + \frac{1}{a_2} + \cdots + \frac{1}{a_n},$$

where n is the first integer for which $\left\{ \dfrac{1}{x_n} \right\} = 0$, and then $a_n = \dfrac{1}{x_n}$.

Find this expansion, an example of an "Egyptian Fraction" for $x = \dfrac{11}{13}$.

Answer

$\dfrac{11}{13} = \dfrac{1}{2} + \dfrac{1}{3} + \dfrac{1}{78}$

Solution

We have

$$a_1 = \left\lfloor \frac{13}{11} \right\rfloor + 1 = 2 \Rightarrow x_2 = \frac{11}{13} - \frac{1}{2} = \frac{9}{26}.$$

Then

$$a_2 = \left\lfloor \frac{26}{9} \right\rfloor + 1 = 3 \Rightarrow x_3 = \frac{9}{26} - \frac{1}{3} = \frac{1}{78}.$$

Hence $a_3 = 78$ and we have

$$\frac{11}{13} = \frac{1}{2} + \frac{1}{3} + \frac{1}{78}.$$

Problem 9.8 Let x and the sequences $\{a_k\}$, and $\{x_k\}$ be as in the previous question.

(a) Suppose $x_k = \dfrac{r}{s}$ and $x_{k+1} = \dfrac{t}{u}$ (where both fractions are reduced). Prove that $0 < t < r$.

Solution

First let $n = \left\lfloor \dfrac{s}{r} \right\rfloor$ so we have

$$a_k = \left\lfloor \frac{1}{x_k} \right\rfloor + 1 = \left\lfloor \frac{s}{r} \right\rfloor + 1 = n+1.$$

Thus

$$x_{k+1} = \frac{r}{s} - \frac{1}{n+1} = \frac{rn+r-s}{s(n+1)} = \frac{r+(rn-s)}{s(n+1)}.$$

Since $n = \left\lfloor \dfrac{s}{r} \right\rfloor$ we have that $-r < rn - s < 0$ so $r + (rn - s) < r$. Therefore $t < r$ (as either $t = r + (rn - s) < r$ or the fraction can be further simplified, but again $t < r$).

(b) Prove that the expansion described always produces a finite Egyptian Fraction.

Solution

Note if at any point we have that the numerator of x_k is 1, then $\left\{ \dfrac{1}{x_k} \right\} = 0$ and the sequence terminates.

Further, by part (a), the numerator of the x_k's is strictly decreasing, so one of the numerators is eventually 1, as needed.

Problem 9.9 Solve $x + 2\{x\} = 2\lfloor x \rfloor$.

Answer

$0, 4/3, 8/3$

Solution

Rewrite the equation as $\lfloor x \rfloor + \{x\} + 2\{x\} = 2\lfloor x \rfloor$, so $3\{x\} = \lfloor x \rfloor$. Since $0 \le \{x\} < 1$, and $\lfloor x \rfloor$ is an integer, we get $\{x\} = 0, 1/3, 2/3$, and $\lfloor x \rfloor = 0, 1, 2$. Therefore $x = 0, 4/3, 8/3$.

Problem 9.10 $4x^2 - 40\lfloor x \rfloor + 51 = 0$.

Answer

$\sqrt{29}/2, \sqrt{189}/2, \sqrt{229}/2, \sqrt{269}/2$

Solution

Since $40\lfloor x \rfloor$ is even, $4x^2$ must be odd (remember x is not an integer). Let $4x^2 = 2k + 1$, then $x = \dfrac{\sqrt{2k+1}}{2}$, thus $\left\lfloor \dfrac{\sqrt{2k+1}}{2} \right\rfloor = \dfrac{k+26}{20}$. Since the LHS is an integer, the integer k must be of the form $20m + 14$. Also, $\dfrac{k+26}{20} \le \dfrac{\sqrt{2k+1}}{2} < \dfrac{k+26}{20} + 1$. Simplify and get two equations: $k^2 - 148k + 4 \times 144 \le 0$, and $k^2 - 108k + 24 \times 84 > 0$. Solving these inequalities, $4 \le k < 24$ or $84 < k \le 144$. In these ranges, the k values of the form $20m + 14$ are: $14, 94, 114, 134$. Solve for the corresponding x values to get the final answer.

www.ingramcontent.com/pod-product-compliance
Lightning Source LLC
Chambersburg PA
CBHW081506200326

41518CB00015B/2403